Generationism

Is It Significant?

Mark Roberts-Seymour, PEng., OFS, CD

Published by Secular Orders Press
1018 – 4900 – 20th Street
Vernon, BC, Canada V1T 9W3

Library of Congress Cataloguing:
Mark E. Roberts-Seymour, CD (1948-)
Generationism: Is It Significant?
1. Social Science 2. Demographics, 3. Title

ISBN 9781723807510
Copyright © 2018 by Mark Roberts-Seymour
All Rights Reserved

Abstract

This book considers the development of generation metrics in the Social Sciences. Generations after the First World War are named and considered. Definitions are used to signal trends in income, family and voting across age boundaries. A convention of approximately 20 years for each of the most recent seven 'generations' has been adopted by the majority of Demographers. The rationales for this are reconsidered. The development of 'generational thought' is considered along with the development of criticism of the generationism. A brief approach to statistical method and 'interpretation' is also offered. As Canadian Demographer Norman Ryder professed "grand generational theories tended toward 'arithmetical mysticism."

Table of Contents

Abstract 3
Introduction 6
    Definitions 7
    Current Generational Labelling 8
Development of Generationalism 14
    Karl Mannheim 16
    Generational Discordance 21
    Norman Ryder 21
Life Expectancy and Generationism 24
Criticism of Generationism 26
    Social Implications 28
    Cohorts 29
    Dating the Sample 31
The Use of Statistics 40
    Sample Size 42
Statistical Misuse, Abuse and Ignorance 47
    Misuse 48
    Misinterpretation 51
    Method in Statistics 51
    Census and Selection 54
    Sociology Surveying 56
    Experimental studies 66
    The Hawthorne Effect 68
    Studies 69
Intergenerational Equity 74

Intergenerational Standards of Living 76
Adultism 78
    Underpinnings of Adultism 84
    Stratification 88
    Age of Majority 91
    Youth Rebellion 94
    Secondary Schooling 97
    (Home-based) 97
    Voting 99
Generation-X 102
    Birth Cohorts and US Voting Experience 106
Concluding Remarks 108
Selected Readings 110
About the Author 123

Introduction

History and the influence individuals have imprinted upon it has often been categorized by eras, ages, epochs or centuries. In the last two centuries it has come to reflect upon more or less distinct generations. particularly since the Second World War. Viewing this newer approach and its impact on voter apathy and happiness-contentment is the thrust of this paper.

It is generally accepted that the reactions of the individual to external stimuli are 'predominantly' based on their life experience, which includes their education at home, and since the nineteenth century, in the West, through their educational institutions. Transmitted (and usually major) historical events such as the assassination of John Kennedy, that of Martin Luther King, the Vietnam War, Quebec Referenda and the 9/11 Twin Tower terrorist attack have moulded reactions of successive recent North American 'generations'. When the collective reactions of individuals shape regional and national reactions, group 'outcomes' occur that in turn impact the individuals who have agglomerated – whether the 'field' is popularity of music style or the election of Members of Parliaments. Sourcing and stating the 'positions/snapshots' or 'trending' of groups by cohorts of birth is the purview of Generationism demographers, and specifically of self-styled 'generationalists'.

Definitions

A generation is "all of the people born and living at about the same time, regarded collectively." It can also be described as, "the average period, during which children are born and grow up, become adults, and begin to have children of their own." It is also known as biogenesis.

"Generation" is also often used synonymously with cohort in social science; under this formulation it means "people within a delineated population who experience the same significant events within a given period of time". Generation in this sense of birth cohort, also known as a "social generation", is widely used in popular culture, and is now frequently the basis for sociological analysis. Serious analysis of generations began in the nineteenth century, emerging from an increasing awareness of the possibility of permanent social change and the idea of youthful rebellion against the established social order. Most practicing analysts believe that a generation is one of the fundamental social categories in a society, while fewer view its importance as a flagship of research being overshadowed by the 'traditional' factors such as class, gender, race, educational attainment, along with others.

Social generations are cohorts of people who were born in the same date range and share similar cultural experiences.

Current Generational Labelling

For the purposes of a list, "Western Cultures" can be taken to include North America, Western Europe, Australia and New Zealand. However, it should also be noted that many variations may exist within the regions, both geographically and culturally, which means that the list is broadly indicative, but necessarily very general.

- The Lost Generation is a term originating with Gertrude Stein to describe those who fought in World War I. The members of the lost generation were typically born between 1883 and 1900. (17 years)
- The G.I. Generation is the generation that includes the veterans who fought in World War II. They were born from around 1901 to 1924 (23 years),
- The Silent Generation, also known as the Lucky Few, was born from approximately 1925 to 1942 (17 years). It includes some

who fought in World War II, most of those who fought the Korean War and many during the Vietnam War.
- The Baby Boomers are the generation that was born following World War II. There are no precise dates when the cohort birth years start and end. Typically, they range from 1944 ending somewhere between 1960 and 1965. Increased birth rates were observed during the post–World War II baby boom, making them a relatively large demographic cohort.
- Generation X, is the generation which follows on after the Baby-Boomers as Demographers define it they were born between 1961–1981 (20 years). In a 2012 article for the Joint Center for Housing Studies of Harvard University, George Masnick wrote that the "Census counted 82.1 million" Gen Xers in the U.S., which represents 27% of the gross national population and approximately 39% of the employable fraction."
- Millennials, also known as the Millennial Generation or Generation Y, are the demographic cohort following Generation X. Demographers and researchers typically use 1982-2002 (20 years).
- Generation Z, also known as the Post-Millennials or "the Information. Generation",

is the cohort of people born after the Millennials. Demographers and researchers typically use starting birth years ranging from 2002, while there is little consensus yet regarding ending birth years. Generation Z is known as the first generation not to have experienced life without information and communications technology and social media.

An example of using ageism [generationism] to quantify behaviour is the employment of Cellphone and Smartphones by Americans. The latest available (reliable) figures show 63% of all people in the world are using cellphones (2016). The American proliferation of cellphones has exploded by more than 100% in seven years according the Pew Research Group, leading to the following 2018 distribution of cell and 'smart' phones:

Age of User	Any Cellphone % of Population	Smartphones % of Population
All	95	85
Age 18-29	100	94
Age 30-49	98	89
Age 50-64	94	74

| Age 65+ | 85 | 46 |

Using the artificial nomenclature for 'labelled 'recent generations published data observed that Millennial and Generation X populations are less likely to vote than are Baby Boomer Generation populations, both in Canada and the US:

- . Among Millennials there is a US rate of 41% voting while the equivalent Canadian rate is 57% - 39% more within cohort.
- For Generation X the rates increase to 57% (US) and 67% (Canada) respectivelly.
- For Baby Boomers rates of voting are higher than either of the other later generations.

In Australia 100% participation in voting is required of all generations of eligible voters (those male and female born before 1999). Failure to vote without legitimate excuse is a 'crime; in that country.

It is suggested in the literature that Generation X persons are self-identified as happy-contented with their lot, and accordingly do not especially engage themselves with social activism directed toward 'change' (as would be evidenced by exercise of voting franchise). This is interpreted by many as apathy born of contentment with their lot, the status-quo and conservatism. It is further 'interpreted' that this generation is hard-working to maintain their desired lifestyle, the rearing of children, meet significant debt and support parents. They change employers more frequently than prior generations to 'get ahead'. They also are more likely to act as entrepreneurs. Generation X workers are likely to be two partner working households as their individual incomes are considerably lower at the same ages than was the equivalent case for the preceding Baby Boomer Generation. They are said to also likely have more children and extended families reliant upon them.

However, caution is in order. Canadian Sociologist Norman Ryder railed at theorists he called 'generationists'. He argued that thinkers about generation on a large scale had made illogical leaps when theorising the relationship between generations and social change. 'The fact that social change produces intercohort differentiation and thus contributes to inter-generational conflict,' he argued, 'cannot justify a theory that social change is produced by that conflict.' There was no way to prove causality. The end result, he wrote, was that grand generational theories tended toward 'arithmetical mysticism.'

In the text which follows this author will address some of the 'offerings' by generationists together with an examination of pitfalls created through both labelling and statistically interpreting generationally.

It is very reasonable to ask of any generalization [e.g. Millennials are prone to feel entitled] to ask who what where when how and why:

2. Who says so [The Observer], and who funded the study?
3. Who are the Millennials surveyed [e.g. rich, poor, college graduate or not]?
4. What is the 'quality' of the study?
5. Where did the sampling take place [e.g. all of North America, the Western Hemisphere etc.]?
6. When was this study done [and is it relevant today]?
7. How was the study conducted?
8. Why was this particular study conducted?

Development of Generationalism

The idea of a social generation, in the sense that it is used today, gained currency in the early 19th century. Earlier concepts of "generation" had referred to family relationships and not broader social groupings. Several trends promoted a new idea of generations, and as the 19th century progressed, of a society divided into different categories of people based on age. These trends were all related to the four processes of:

1. modernisation,
2. education,
3. industrialisation, or
4. westernisation,

which had been changing the face of Western Cultures since the mid-18th century. One was a change in mentality about time and social change. Enlightenment ideas encouraged the idea that society and life were changeable, and that civilisation could progress. This promulgated an equating of youth with social renewal and change. Political rhetoric often focused on the renewing power of youth, giving rise to 'youth groups' with nationalist flavours, (i.e. Young Italy, Young Germany). By the end of the 19th century,

Western intellectuals were inclined to think of the world in generational terms—in terms of both youth rebellion and emancipation.

An important contributing factor resulting in a change was the change in the economic structure of society. Because of the rapid social and economic change, young men [particularly] were less beholden to their fathers and family authority than they had been, though they were frequently beholden to national military service. Greater social and economic mobility allowed them to assert their authority to a much greater extent than had traditionally been possible. The skills and wisdom of fathers were often less valuable than they had been due to technological and social change. During this time, the period between childhood and adulthood, usually spent at education or in military service, was also increased for many people entering 'sedentary jobs'. This category of people was very influential in spreading the ideas of youth culture renewal.

The breakdown of traditional social and regional identifications is noteworthy. The spread of nationalism and many of the factors that created it:

- a national press,
- linguistic homogenisation,

- public education,
- suppression of local particularities,
- suppression of religious authority

encouraged a broader sense of belonging beyond local affiliations. Individuals thought of themselves increasingly as part of a larger 'society', and this encouraged identification with groups beyond the local. Auguste Comte (1798-1857) was the first philosopher to make a serious attempt to systematically study generations. In *Cours de philosophie positive* Comte suggested that social change is determined by generational change; and in particular conflict between successive generations. As the members of a given generation 'mature', their "instinct of social conservation" becomes stronger, which inevitably and necessarily brings them into conflict with the "normal attribute of youth"—innovation. Other important social-generational theorists of the 19th century were John Stuart Mill (1806-1873) and Wilhelm Dilthey (1833-1911).

Karl Mannheim

Sociologist Karl Mannheim (1893-1947) was a seminal figure in the study of generations. He elaborated a theory of generations in his 1923 essay *The Problem of Generations*. He suggested that there had been a division into two primary schools of study of generations until that time. Firstly, positivists such as Comte measured social change in specific life-spans. Mannheim argued that this reduced history to "a chronological table". The other school, the "romantic-historical" was represented by Dilthey and Martin Heidegger (1889-1976). This school focused on the individual qualitative experience at the expense of social context.

Mannheim, born in Budapest in 1893, had a life marked by 20th-century political upheaval. He was a student during the First World War, had to leave Hungary for Germany when the Kingdom of Hungary was reinstated in 1920, then had to leave Germany for England when the Nazi regime came to power and he lost his professorship for being Jewish. Despite the huge impact that the historical coincidence of his birth had on his life, Mannheim wrote his 1928 essay in order to temper enthusiasm about the broad generational explanations that were *en vogue* in the European intellectual community.

It was time, Mannheim wrote, to think more systematically about this attractive way to explain historical change. He wrote that the logical fallacy of the generational approach 'lay in the attempt to explain the whole dynamic of history from this one factor – an excusable one-sidedness easily explained by the fact that discoverers often tend to be over-enthusiastic about phenomena they are the first to see'.

Contemporaries and successors to Mannheim identified Positivist thinkers – mostly French [influenced by the philosopher Auguste Comte] who undertook to 'find a general law to express the rhythm of historical development', in keeping with their project of understanding society empirically in order to rationally direct its course. Looking at the average life course of humans, these thinkers and writers tied the progress – or lack thereof – of human culture directly to this biological limitation, and wondered how things would change if humans lived longer or shorter lives. In the 21st century we are faced with a protraction [increase in the life expectancy of each newly born] 'generations'.

Mannheim emphasised that the rapidity of social change in youth was crucial to the formation of generations, and that not every generation would come to see itself as distinct. In periods of rapid social change, a generation would be much more likely to develop a cohesive character. He also believed that a number of distinct sub-generations could exist, and these would have shared unique experience-perception. Mannheim identified three commonalities that a generation share:

- Shared temporal location – generational site or birth cohort
- Shared historical location – generation as actuality or exposure to a common era
- Shared sociocultural location – generational consciousness or "entelechy" (A particular type of motivation, need for self-determination, and inner strength directing life and growth to become all the individual is capable of being; the need to actualize one's beliefs; having both a personal vision and the ability to actualize that vision from within.)

Karl Mannheim, in his 1952 book *Essays on the Sociology of Knowledge* asserted the belief that people are shaped through lived experiences as a result of social change. Based on the way lived experiences shape a generation in regard to values, the result is that the new generation will challenge the older generation's values, resulting in tension. This challenge between generations and the tension that arises is a defining point for understanding generations and what separates them.

The concept of a generation is also used to locate particular birth cohorts in specific historical and cultural circumstances, such as the "Baby boomers", a nomenclature drawn from the high birth rate during that period.

While all generations have similarities, there are differences among them as well. A 2007 report *"Millennials: Confident. Connected. Open to Change"* noted the challenge of studying generations: "Generational analysis has a long and distinguished place in social science, and we cast our lot with those scholars who believe it is not only possible, but often highly illuminating, to search for the unique and distinctive characteristics of any given age group of North Americans. But we also know this is not an exact science. It can be argued it is not a science at all. We are mindful that there are as many differences in attitudes, values, behaviours, and lifestyles within a generation, as there are between generations. But we believe this reality does not diminish the value of generational analysis; it merely adds to its richness and complexity."

Another element of generational theory is recognizing how youth experience their generation, and how that changes based on where they reside in the world. "Analyzing young people's experiences in place contributes to a deeper understanding of the processes of individualization, inequality, and of generation."

Generationism

Being able to take a closer look at youth cultures and subcultures in different times and in different places adds an extra element to understanding the everyday lives of youth. This allows a better understanding of the ways generation (and place) affects development.

It is not where the birth cohort boundaries are drawn that is most significant, but the manner in which individuals and societies interpret labelled boundaries and how the artificial divisions shape processes and outcomes. Unfortunately, the practice of categorizing age cohorts has permitted researchers to construct.

Generational Discordance

Norman Ryder

Canadian Demographer and Social Scientist Norman Ryder (1924-2009), in 1965, approached the sociology of discord between generations by suggesting that society "persists despite the mortality of its individual members, through processes of demographic metabolism and particularly the annual infusion of birth cohorts". He argued that generations may sometimes be a "threat to stability" but at the same time they represent "the opportunity for social transformation". Ryder attempted to understand the dynamics at work across generations.

Amanda Grenier, in a 2007 essay [Journal of Social Issues], offered a contrasting explanation for why generational tensions exist. She asserted that generations develop their own linguistic models that contribute to misunderstanding between age cohorts, "Different ways of speaking exercised by older and younger people exist, and may be partially explained by social historical reference points, culturally determined experiences, and individual interpretations". It is also noteworthy that linguistic character is directly affected by the nature of the educational environment experienced by successive generations. Language taught (or 'de-emphasised') is critical to the manner in which an environment is perceived by individuals and groups. What is 'known' by a cohort is largely dependent on the facility with language that they can manipulate.

Writers began creating theories about generational meaning in the 19th century. By the time the Hungarian-born sociologist Karl Mannheim wrote the formative essay 'The Problem of Generations' (1928), he could include 33 references in his bibliography, from European scholars who had been addressing the question since 1862. A pervasive 18th- and 19th-century European intellectual interest in modernisation, progress and change laid the groundwork for assessments of age groups as separate entities whose efforts would bring societies closer, or distance them, from the goals that writers set for their cultures. Then, traumatic and socially significant experiences of revolution and war in the 19th and 20th centuries disposed people to try to understand how such dramatic changes could happen so quickly – and how they might change those who lived through them. Young people began to self-define as fundamentally different from older people, and to take political action based on those generational beliefs.

Life Expectancy and Generationism

There is a significant shift in the life expectancy of succeeding generations since the first World War. At the time of Mannheim's seminal works the influence of the then current life cycle of the individual stood at 58 year for males and 62 years for females [as noted in the graph following. By 2015 the life expectancy had increased to 78 year for men and 83 years for females. It is reasonable to assume that the social contribution of any generation has accordingly changed simply because they live longer – and that is critical to reflection on 'patterns of behaviour' by labelled generations.

Life expectancy at birth, by sex, Canada, 1956 to 2005

Source: 1921 to 1981: Nagnur D. Longevity and Historical Life Tables, 1921 to 1981 (Abridged), Statistics Canada, Catalogue 98-506, 1986;
1986: Duchesne D, Nault F, Gilmour H, Wilkins R. Vital Statistics Compendium 1996, Statistics Canada, Catalogue 84-214, 1999;
1991 to 2005: CANSIM Table 102-0511, Life expectancy, abridged life table, at birth and at age 65, by sex, Canada, provinces and territories, annual.

Generationism

From 1950 to 2014 the world life expectancy jumped by fifty percent. Projections vary considerably: but before 2050 it is quite likely to exceed 100 years (it already does for those now living to age 90, if in good health). The following table indicates some benchmark life spans divided by era:

Era	Life expectancy at birth in years	Threshold % of Canadian 2017 LEB
Paleolithic	33	
Neolithic	20 to 33	41.3
Bronze Age and Iron Age	26	
Classical Greece	25 to 28	35.0
Classical Rome	20–30	37.5
Pre-Columbian Southern United States	25–30	37.5
Medieval Islamic Caliphate[	35+	
Late medieval English peerage	30	
Early modern England	33–40	50.0
Pre-Champlain Canadian Maritimes	60	75.0
18th-century Prussia	24.7	30.9

Generationism

Era	Life expectancy at birth in years	Threshold % of Canadian 2017 LEB
18th-century France	27.5–30	37.5
18th-century Qing China	39.6	49.5
18th-century Edo Japan	41.1	
Early 19th-century England	40	51.4
1900 world average	31	38.8
1950 world average	48	60.0
2014 world average	71.5	89.4

Criticism of Generationism

Today there is a booming market for pundits who can devise grand theories of generational difference. Neil Howe and William Strauss, authors of *Generations: The History of America's Future, 1584-2069* (1991) and founders of the consulting firm LifeCourse Associates in Virginia, have made a fine living out of generational assessments, but their work reads like a deeply mystical form of historical explanation. These two authorities conceived an elaborate and totalising theory of the cycle of generations, which they argue come in four sequential and endlessly repeating archetypes. The distinct groups of archetypes follow each other throughout history thus: 'prophets' are born near the end of a 'crisis'; 'nomads' are born during an 'awakening'; 'heroes' are born after an 'awakening', during an 'unravelling'; and 'artists' are born after an 'unravelling', during a 'crisis'.

> 'prophets' post-crisis>
> 'nomads' during 'awakening'>
> 'heroes' post 'awakening', during an 'unravelling'>
> artists' after an 'unravelling', during a 'crisis'

The archetypal scheme is also a theory of how historical change happens. The LifeCourse idea is that the predominance of each archetype in a given generation triggers the advent of the next (as the consultancy's website puts it: 'each youth generation tries to correct or compensate for what it perceives as the excesses of the midlife generation in power'). Besides having a very reductive vision of the universality of human nature, Strauss and Howe are futurists; they predict that a major crisis will occur once every 80 years, restarting the generational cycle. While the pair's ideas seem far-fetched, they have currency in the marketplace: LifeCourse Associates acts for brands such as Nike, Cartoon Network, Viacom and the Ford Motor Company; for universities including Arizona State, Dartmouth, Georgetown and the University of Texas, and for the US Army, too.

Social Implications

A majority of North American thinkers on the generational question tend to flatten social distinctions, relying on hand-picked examples and positing a vision of a 'society' that's made up mostly of the white and middle-class. Eric Hoover offered a withering perspective of Howe and Strauss's influential book about that generation, *Millennials Rising: The Next Great Generation* (2000), as' a work 'based on a hodgepodge of anecdotes, statistics, and pop-culture references' with the only new empirical evidence being a body of around 600 interviews of high-school seniors, all living in wealthy Fairfax County, Virginia. Siva Vaidhyanathan, a cultural historian and media scholar at the University of Virginia, told Hoover: 'Generational thinking is just a benign form of bigotry.'

Academics have been working and re-working the concept of 'generations' for more than a century, and have generally concluded that generational thinking is bogus. Distinctions between given age groups in a society can be an interesting lens for examination – but only if the person framing the questions is painfully cautious to qualify her terms, set careful parameters, and examine her assumptions.

No available models totally satisfied Mannheim, who felt that biological (positivist) or spiritual (romantic) speculations about the nature of 'secular rhythms' in history should be advanced with caution. These speculations, he thought, were 'simply used as a pretext for avoiding research into the nearer and more transparent fabric of social processes and their influence on the phenomenon of generations'. And no set interval of generational spread – 30 years; 15 years – should be accepted as gospel, since intermediate generations always played a part in the development of the generations around them.

Cohorts

Mannheim cautioned recognition of the existence of diverse 'generation units' [viz. youth of Fairfax County and young people living in the Rio Grande Valley in Texas or the South Side in Chicago] – was essential. As evidence, he pointed to European peasants living outside of cities in the 18th and 19th centuries, who couldn't possibly have the same perspective as their urban, educated brethren on the upheavals of revolution. It was tempting, he admitted, to make literary or artistic groups stand in for the rest of their generation, since such self-reflective, highly analytical groups made entelechies really visible. 'But if we pay exclusive attention to them,' he warned, 'we shall not be able really to account for this vector structure of intellectual currents.'

The Canadian sociologist Norman Ryder redefined the problem with *The Cohort as a Concept in the Study of Social Change'* (1965), which is still widely cited in sociological literature dealing with age, life course and experience. 'A cohort may be defined as the aggregate of individuals (within some population definition) who experienced the same event within the same time interval,' Ryder argued. That qualifier – 'within some population definition' – was key. Sociologists looking at cohorts could avoid oversimplifying their data by always controlling for other factors relating to social position: geographical location, gender, race, education, occupation. More often than not they failed to make this move.

Making assumptions concerning a given group of people would be similar because of birthdate, Ryder thought, was to risk committing a fallacy. 'The burden of proof is on those who insist that the cohort acquires the organised characteristics of some kind of temporal community,' he wrote. 'This may be a fruitful hypothesis in the study of small groups of coevals in artistic or political movements but it scarcely applies to more than a small minority of the cohort in a mass society.'

Ryder railed at the theorists he called 'generationists'. He argued that thinkers about generation on a large scale had made illogical leaps when theorising the relationship between generations and social change. 'The fact that social change produces intercohort differentiation and thus contributes to inter-generational conflict,' he argued, 'cannot justify a theory that social change is produced by that conflict.' There was no way to prove causality. The end result, he wrote, was that grand generational theories tended toward 'arithmetical mysticism.'

Dating the Sample

Generationism

When sociologic traits are sampled the date of sampling is crucial. Baby-Boomers, considered by many today to be archly conservative [when contrasted with the later generations], were the instigators of the Free Speech Movement, Anti-Vietnam War protestations, The Civil Rights demonstrations, and A Unique Drug Culture along with a myriad of other reactionary liberal undertakings. This points to the importance of the dating [benchmarking] when data concerning each generation is obtained. In 1968 the generation's internal 'appearances' would not fit with those same persons sampled in 2018.

When studies are made there is strong evidence that the date of sampling should be carefully scrutinized, and that repetition at an interval [or intervals] should be incorporated in the foundation of the initial grant.

While sociologists of the past half-century have used Ryder's cohort concept to delve into human experience – structuring their studies using the variable of birthdate along with any number of other defining facts that might shape human lives – some historians have also tried to recover the generational idea by writing about specific cohorts. By mid-century, cadres of historians, reflecting on their own methodology, saw earlier efforts to propose long-term, large-scale generational schemes as hopelessly arbitrary and potentially deeply flawed, [e.g. François Mentré, Henri Peyre, and Julián Marías].

The US historian Robert Wohl wrote, in his book *Generation of 1914* (1979), about the European intellectuals who self-defined as a group after the Great War. In wrestling with the idea of 'generation', Wohl looked at the self-conscious creation of the concept in post-war Europe. Wohl's key theoretical move was to skirt the question of the existence of a generation, instead making a meta-analysis, looking at the way that cohorts of intellectuals, artists and academics developed their specific generational consciousness. Wohl studied the formation of the idea, rather than the actuality.

Wohl painstakingly detailed the differences between units within this 'generation', even those separated by only a few years. Slightly older men, who entered the First World War early in the conflict, after establishing nascent careers and while the project of the war was still seen as an honourable one, had a very different experience from those younger soldiers who went straight from school to war. While both groups were later considered part of the 'Lost Generation' of British elites, Wohl wrote, the survivors' social position and attitudes upon their re-entry after the war were quite different

The people who made up Wohl's 'Generation of 1914' were said to share many similar intellectual obsessions, including:

1. the casting-off of traditions,
2. an emphasis on action,
3. a prizing of authenticity
4. avant-garde thought.

Wohl wrote, "the survivors' social position and attitudes upon their re-entry after the war were quite different. Biological (positivist) or spiritual (romantic) speculations about the nature of 'secular rhythms' in history should be advanced with caution." These speculations, he thought, were 'simply used as a pretext for avoiding research into the nearer and more transparent fabric of social processes and their influence on the phenomenon of generations'. No set interval of generational spread – 30 years; 15 years – should be adopted, since intermediate generations always played a part in the development of the generations before and after any selected arbitrary dates.

For these purposes, Mannheim cautioned recognition of the existence of diverse 'generation units' [e.g. between the present-day youth of Fairfax County and young people living in the Rio Grande Valley in Texas or the South Side in Chicago] was essential. As evidence, he considered European peasants living outside of cities in the 18th and 19th centuries, who couldn't possibly have the same perspective as their urban, educated brethren on the upheavals of revolution. It was tempting, he admitted, to make literary or artistic groups stand in for the rest of their generation, such clusters of self-reflective, highly analytical groups made vital comparisons really visible. 'But if we pay exclusive attention to them,' he warned, 'we shall not be able really to account for this vector structure of intellectual currents.'

Norman Ryder redefined the problem with *'The Cohort as a Concept in the Study of Social Change'* (1965), which continues to be widely cited in sociological literature dealing with age, life course and experience. 'A cohort may be defined as the aggregate of individuals (within some population definition) who experienced the same event within the same time interval,' Ryder argued. That qualifier – 'within some population definition' – was key. Sociologists looking at cohorts could avoid oversimplifying their data by always controlling for other factors relating to social position: geographical location, gender, race, education, occupation.

To assume that a given group of people would be similar because of birthdate, Ryder thought, was to risk committing a fallacy. 'The burden of proof is on those who insist that the cohort acquires the organised characteristics of some kind of temporal community,' he wrote. 'This may be a fruitful hypothesis in the study of small groups of coevals in artistic or political movements but it scarcely applies to more than a small minority of the cohort in a mass society.'

Ryder railed at theorists he called 'generationists'. He argued that thinkers about generation on a large scale had made illogical leaps when theorising the relationship between generations and social change. 'The fact that social change produces intercohort differentiation and thus contributes to inter-generational conflict,' he argued, 'cannot justify a theory that social change is produced by that conflict.' There was no way to prove causality. The end result, he wrote, was that grand generational theories tended toward 'arithmetical mysticism.'

While sociologists of the past half-century have used Ryder's cohort concept to see deeper into human experience – structuring their studies using the variable of birthdate along with any number of other defining facts that might shape human lives – some historians have also tried to recover the generational idea by writing about specific cohorts. By mid-century, historians reflecting on their own methodology saw earlier efforts to propose long-term, large-scale generational schemata, such as those of François Mentré, Henri Peyre, and Julián Marías, as hopelessly arbitrary.

Currently, some historians have tried to think generationally while rigorously acknowledging the structural limitations of the approach. In her book *Inheriting the Revolution: The First Generation of Americans* (2000), US historian Joyce Appleby considers the generation that came of age between 1790 and 1830, and was fundamentally shaped by the experience of being the first adults to grow up in the new nation. She posits that religious revival, economic opportunity and democratic politics made their mark on these people, who created their own ideology around what it meant to be a citizen.

Pierre Nora wrote in 1996: "the careful analyst trying to talk about generations will always struggle": 'The generational concept would make a wonderfully precise instrument if only its precision didn't make it impossible to apply to the unclassifiable disorder of reality.' The problem with transferring historical and sociological ways of thinking about generational change into the public sphere is that 'unclassifiability' is both terrifying and boring. It is unfortunate that big, sweeping explanations of social change sell when made available for public consumption. It is also, as a corollary the way to induce funders to sponsored subsequent studies, and academia bristles with those who jump onto the bandwagon to secure new study grants. Careful studies of same-age cohorts, hemmed in on all sides by rich specificity, do not sell.

The presumption of a "science" of supposed 'generations' would chafe less if it weren't so often used to demean the young. Millennials. 'Consultants' advise prospective employers, that Millennials feel entitled to good treatment even in entry-level jobs, because they've been overpraised their whole lives. 'Millennials won't buckle down and buy cars or houses, economists complain; millennials are lurking in their parents' basements, tweeting and texting and posting selfies – and avoiding responsibility'. It is far more accurate to say that millennials are fighting back, pointing out that this focus on technology use and supposed personality differences is obscuring the very real (and dire) economic conditions that young people face. 'Sometimes it's important to start with numbers,' Malcolm Harris wrote in the 'Youth' edition of *The New Inquiry* in 2012. 'When it comes to inter-generational conflict, tied as it is to stories about Oedipus and Hamlet, numbers help ensure we're speaking of a particular relation rather than a mythic archetype.' Harris has remarked that young people deal with:

1. unemployment,
2. overpolicing,
3. lack of economic opportunity,
4. tuition increases, and
5. mountainous student debt.

The Use of Statistics

It has been suggested above that cohorts form a critical part of interpretation of data concerning generationism. It goes ultimately to statistical inference, and particularly to three other issues:

- bias of the observer and study formulator(s),
- randomness [where whole populations are not sampled] and
- population [N size].

Statistics is math which deals with data:

1. collection,
2. organisation,
3. analysis,
4. interpretation and
5. presentation

In applying statistics to a social issue such as 'generational traits', it is conventional to begin with a statistical population or a statistical model process to be studied. Populations can treat diverse. topics such as "all people living in a country". Statistics deals with all aspects of data including the planning of data collection in terms of the design of surveys and experiments. When census data cannot be used, analysts collect data by developing specific experiment designs and survey samples. Representative sampling [when it actually occurs with respect to funding] assures that inferences and conclusions can reasonably extend from the sample to the population as a whole. Critically the 'experimental study' involves taking measurements of the system under study, manipulating the system, and then taking additional measurements using the same procedure to determine if the manipulation has modified the values of the measurements. In contrast, an 'observational study' does not involve experimental manipulation. This latter is the form most employed in generational inference and is prone to error.

The rapid and sustained increases in computing power starting from the second half of the 20th century have had a substantial impact on the practice of statistical science. Early statistical models were almost always from the class of linear models, but powerful computers, coupled with suitable numerical algorithms, caused an increased interest in nonlinear models (such as neural networks) as well as the creation of new types, such as generalized linear models and multilevel models.

Increased computing power has also led to the growing popularity of computationally intensive methods based on resampling, while techniques such as Gibbs sampling have made use of Bayesian models more feasible. The computer revolution has implications for the future of statistics with new emphasis on "experimental" and "empirical" statistics. A large number of both general and special purpose statistical software are now available. Examples of available software capable of complex statistical computation include programmes such as Mathematica, SAS, SPSS, and R.

Sample Size

Sample size (N) determination involves choosing the number of observations to be included in a statistical sample. The sample size is an important feature of any empirical study in which the goal is to make inferences about a population from a sample. In practice, the sample size used in a study is determined based on the expense of data collection, and the need to have sufficient statistical power. In complicated studies there may be several different sample sizes involved in the study: for example, in a stratified survey there would be different sample sizes for each stratum. In a census, data are collected on the entire population, hence the sample size is equal to the population size.

Sample sizes may be chosen in several different ways:

- experience – A choice of small sample sizes, though sometimes necessary, can result in wide confidence intervals or risks of errors in statistical hypothesis testing.
- using a target variance for an estimate to be derived from the sample eventually obtained, i.e. if a high precision is required (narrow confidence interval) this translates to a low target variance of the estimator.
- using a target for the power of a statistical test to be applied once the sample is collected.
- using a confidence level, i.e. the larger the required confidence level, the larger the sample size (given a constant precision requirement).

Larger sample sizes generally lead to increased precision when estimating unknown parameters.

Generationism

For example, if we wish to know the proportion of 20-year-old persons who own smartphones it would generally produce a more precise estimate of this proportion if we sampled and examined 200 rather than 100 persons. Several fundamental facts of mathematical statistics describe this phenomenon, including the law of large numbers and the central limit theorem. [CLT] establishes that, in some situations, when independent random variables are added, their properly normalized sum tends toward a normal distribution (informally a "*bell curve*") even if the original variables themselves are not normally distributed. The theorem is a key concept in probability theory because it implies that probabilistic and statistical methods that work for normal distributions can be applicable to many problems involving other types of distributions.

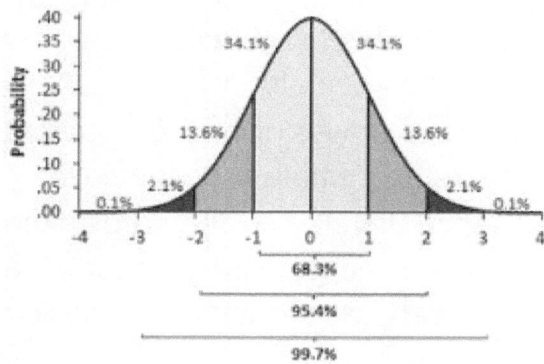

If a sample is obtained containing a large number of observations, each observation being randomly generated in a way that does not depend on the values of the other observations, and that the arithmetic average of the observed values is computed.

When this procedure is performed many times, the central limit theorem asserts that the computed values of the average will be distributed according to a normal distribution. A simple example of this is that if one flips a coin many times the probability of getting a given number of heads in a series of flips will approach a normal curve, with mean equal to half the total number of flips in each series. (In the limit of an infinite number of flips, it will equal a normal curve.)

The central limit theorem has a number of variants. In its common form, the random variables must be identically distributed. In variants, convergence of the mean to the normal distribution also occurs. The earliest version of this theorem, that the normal distribution may be used as an approximation to the binomial distribution, is now known as the de Moivre–Laplace theorem.

In more general usage, a central limit theorem is any of a set of weak-convergence theorems in probability theory. They all express the fact that a sum of many independent and identically distributed random variables, or alternatively, random variables with specific types of dependence, will tend to be distributed according to one of a small set of *attractor distributions*. When the variance of variables is finite, the attractor distribution is the normal distribution.

In some situations, the increase in precision for larger sample sizes is minimal, or even non-existent. This can result from the presence of systematic errors or strong dependence in the data, or if the data follows a heavy-tailed distribution.

Sample sizes are judged based on the quality of the resulting estimates. For example, if a proportion is being estimated, one may wish to have the 95% confidence interval be less than 0.06 units wide. Alternatively, sample size may be assessed based on the power of a hypothesis test. For example, if we are comparing the support for a certain political candidate among women with the support for that candidate among men, we may wish to have 80% power to detect a difference in the support levels of 0.04 units.

Statistical Misuse, Abuse and Ignorance

This author has often cited an example to define incredulous interpretation of stats:

> "It is statistically correct to state that children with bigger shoe sizes are smarter than children with smaller shoe sizes". ´

In real world terms this statement has glossed over the fact that children with bigger shoes sizes are generally 'older' than children with smaller shoe sizes [and that they have accordingly a greater breadth of knowledge-experience, hence they are probably 'smarter' than the cohort of younger children]. Accordingly, while the statement of confidence in the data [shoe size] may be 'true', it is of no practical value. The statistician arriving at an interpretation missed-the-boat. Knowing [with greater skill] that older children are smarter than younger children is not worthy of a real-world study to begin with, not to mention that someone has to pay for such a 'study'.

Generationism

Other 'statements of fact' that bear on the popular misconceptions of generational 'traits' are equally misleading [viz. millennials can't keep a job (in which the choice to change employers is more common today for those attempting to advance more rapidly within their work field and accrue more wealth as a result of job-hopping) when compared to long-service with one employer].

Misuse

Misuse of statistics can produce subtle, but serious errors in description and interpretation—subtle in the sense that even experienced professionals make such errors, and serious in the sense that they can lead to devastating decision errors. For instance, social policy, medical practice, and the reliability of structures like bridges all rely on the proper use of statistics. When statistical techniques are correctly applied, the results can be difficult to interpret for that lacking expertise. The statistical significance of a trend in the data—which measures the extent to which a trend could be caused by random variation in the sample—may or may not agree with an intuitive sense of its significance. The set of basic statistical skills (and skepticism) that people need to deal with information in their everyday lives properly is referred to as statistical literacy.

There is a general perception that statistical knowledge is all-too-frequently intentionally misused by finding ways to interpret only the data that are favorable to the presenter. A mistrust and misunderstanding of statistics is associated with the quotation, "There are three kinds of lies: lies, damned lies, and statistics". Misuse of statistics can be both inadvertent and intentional, and the book *How to Lie with Statistics* outlines a range of considerations. In an attempt to shed light on the use and misuse of statistics, reviews of statistical techniques used in particular fields are conducted (e.g. Warne, Lazo, Ramos, and Ritter (2012)).

To avoid misuse of statistics the works must use proper diagrams and avoid bias. Misuse can occur when conclusions are overgeneralized and claimed to be representative of 'more than they really are', often by either deliberately or unconsciously overlooking sampling bias. Bar graphs are arguably the easiest diagrams to use and understand, and they can be made either by hand or with simple computer programmes. Unfortunately, most people do not look for bias or errors, so they are not noticed. Thus, people may often believe that something is true even if it is not well represented. To make data gathered from statistics believable and accurate, the sample taken must be representative of the whole. According to Huff, "The dependability of a sample can be destroyed by [bias]... allow yourself some degree of skepticism."

To assist in the understanding of statistics Huff proposed a series of questions to be asked in every case:

1. Who says so? (Does he/she have an axe to grind?)
2. How does he/she know? (Does he/she have the resources to know or establish the facts?)
3. What's missing? (Does he/she give us a complete picture?)
4. Did someone change the subject? (Does he/she offer us the right answer to the wrong problem?)
5. Does it make sense? (Is his/her conclusion logical and consistent with what we already know?)

Every person assailed with statistics to bolster an argument should rightfully ask these five questions, and follow-up with the query 'when precisely was this sampled, and does it remain true or relevant when used in argument'.

Generationism

Misinterpretation

The concept of correlation is particularly noteworthy for the potential confusion it can cause. Statistical analysis of a data set often reveals that two variables (properties) of the population under consideration tend to vary together, as if they were connected. For example, a study of annual income that also looks at age of death might find that poor people tend to have shorter lives than affluent people. The two variables are said to be correlated; however, they may or may not be the cause of one another. The correlation phenomena could be caused by a third, previously unconsidered phenomenon, called a lurking variable or confounding variable. For this reason, there is no way to immediately infer the existence of a causal relationship between the two variables.

Method in Statistics

Two main statistical methods are used in data analysis: descriptive statistics, which summarize data from a sample using indices such as the mean or standard deviation, and inferential statistics, which draw conclusions from data that are subject to random variation (e.g., observational errors, sampling variation).

Descriptive statistics are most often concerned with two sets of properties of a *distribution* (sample or population): *central tendency* (or *location*) seeks to characterize the distribution's central or typical value, while *dispersion* (or *variability*) characterizes the extent to which members of the distribution depart from its center and each other. Inferences on mathematical statistics are made under the framework of probability theory, which deals with the analysis of random phenomena.

A standard statistical procedure involves the test of the relationship between two statistical data sets, or a data set and synthetic data drawn from an idealized model. A hypothesis is proposed for the statistical relationship between the two data sets, and this is compared as an alternative to an idealized null hypothesis of no relationship between two data sets. 'Rejecting or disproving' the null hypothesis is done using statistical tests that quantify the sense in which the null can be proven false, given the data that are used in the test. Working from a null hypothesis, two basic forms of error are recognized:

- Type I errors (null hypothesis is falsely rejected giving a "false positive") and
- Type II errors (null hypothesis fails to be rejected and an actual difference between populations is missed giving a "false negative")

Multiple problems have come to be associated with this framework: ranging from obtaining a sufficient sample size to specifying an adequate null hypothesis

Measurement processes that generate statistical data are also subject to error. Many of these errors are classified as random (noise) or systematic (bias), but other types of errors (e.g., blunder, such as when an analyst reports incorrect information) can also be important. The presence of missing data or censoring may result in biased estimates and specific techniques have been developed to address these problems.

It was not until the 18th century that stats started to draw more heavily from calculus and probability theory. Since the end of the twentieth century statistics has relied more on statistical software to produce tests such as descriptive analysis.

Generationism

Census and Selection

Statisticians usually compile data about the entire population (an operation called census). This may be organized by governmental statistical institutes. Descriptive statistics can be used to summarize the population data. Numerical descriptors include mean and standard deviation for continuous data types (like income), while frequency and percentage are more useful in terms of describing categorical data (like race).

When a census is not feasible, a chosen subset of the population called a sample is studied. Once a sample that is representative of the population is determined, data is collected for the sample members in an observational or experimental setting. Again, descriptive statistics can be used to summarize the sample data. However, the drawing of the sample has been subject to an element of randomness, hence the established numerical descriptors from the sample are also due to uncertainty. To still draw meaningful conclusions about the entire population, inferential statistics is needed. It uses patterns in the sample data to draw inferences about the population represented, accounting for randomness. These inferences may take the form of:

1. answering yes/no questions about the data (hypothesis testing),
2. estimating numerical characteristics of the data (estimation),
3. describing associations within the data (correlation) and
4. modeling relationships within the data (for example, using regression analysis).

Inference can extend to forecasting, prediction and estimation of unobserved values either in or associated with the population being studied; it can include extrapolation and interpolation of time series or spatial data, and can also include data mining.

To use a sample as a 'reflection' of an entire population, it is important that it truly represents the overall population. Representative sampling assures that inferences and conclusions can safely extend from the sample to the population as a whole. A major problem lies in determining the extent that the sample chosen is actually representative. Statistics offers methods to estimate and correct for any bias within the sample and data collection procedures. There are also methods of experimental design for experiments that can lessen these issues at the outset of a study, strengthening its capability to discern truths about the population.

Sampling theory is part of the mathematical discipline of probability theory. Probability is used in mathematical statistics to study the sampling distributions of sample statistics and, more generally, the properties of statistical procedures. The use of any statistical method is valid when the system or population under consideration satisfies the assumptions of the method. The difference in point of view between classic probability theory and sampling theory is, roughly, that probability theory starts from the given parameters of a total population to deduce probabilities that pertain to samples. Statistical inference, however, moves in the opposite direction—inductively inferring from samples to the parameters of a larger or total population.

Sociology Surveying

Researchers carry out surveys with a view towards making statistical inferences about the population being studied, and such inferences depend strongly on the survey questions used.

Generationism

Polls about public opinion, public-health surveys, market-research surveys, government surveys and censuses are all examples of quantitative research that use survey methodology to answer questions about a population. Although censuses do not include a "sample", they do include other aspects of survey methodology, like questionnaires, interviewers, and non-response follow-up techniques. Surveys provide important information for all kinds of public-information and research in sociology

Target populations can range from the general population of a given country to specific groups of people within that country, to a membership list of a professional organisation, or list of students enrolled in a school system (see also sampling (statistics) and survey sampling). The persons replying to a survey are called respondents, and depending on the questions asked their answers may represent themselves as individuals, their households, employers, or other organisation they represent.

Survey methodology as a scientific field seeks to identify principles about the sample design, data collection instruments, statistical adjustment of data, and data processing, and final data analysis that can create systematic and random survey

Generationism

errors. Survey errors are sometimes analyzed in connection with survey cost. Cost constraints are sometimes framed as improving quality within cost constraints, or alternatively, reducing costs for a fixed level of quality. Survey methodology is both a scientific field and a profession, meaning that some professionals in the field focus on survey errors empirically and others design surveys to reduce them. For survey designers, the task involves making a large set of decisions about thousands of individual features of a survey in order to improve it.

The most important methodological challenges of a survey methodologist include making decisions on how to:

- Identify and select potential sample members.
- Contact sampled individuals and collect data from those who are hard to reach (or reluctant to respond)
- Evaluate and test questions.
- Select the mode for posing questions and collecting responses.
- Train and supervise interviewers (if they are involved).
- Check data files for accuracy and internal consistency.

Generationism

- Adjust survey estimates to correct for identified errors.

The aim of any survey is not to describe the sample, but the larger population. This generalizing ability is dependent on the representativeness of the sample, as stated above. Each member of the population is termed an element. There are frequent difficulties one encounters while choosing a representative sample.

One common error that results is selection bias. Selection bias results when the procedures used to select a sample result in over representation or under representation of some significant aspect of the population. For instance, if the population of interest consists of 75% females, and 25% males, and the sample consists of 40% females and 60% males, females are under represented while males are overrepresented.

In order to minimize selection biases, stratified random sampling is often used. This is when the population is divided into sub-populations called strata, and random samples are drawn from each of the strata, or elements are drawn for the sample on a proportional basis.

There are several ways of administering a survey. The choice between administration modes is influenced by several factors, including

1. costs,
2. coverage of the target population,
3. flexibility of asking questions,
4. respondents' willingness to participate and
5. response accuracy.

Different methods create mode effects that change how respondents answer, and different methods have different advantages. The six most common modes of administration can be summarized as:

1. Telephone
2. Mail (post)
3. Online surveys
4. Personal in-home surveys
5. Personal mall or street intercept survey
6. Hybrids of the above.

Questionnaires are the most commonly used tool in survey research. However, the results of a particular survey are worthless if the questionnaire is written inadequately. Questionnaires should produce valid and reliable demographic variable measures and should yield valid and reliable individual disparities that self-report scales generate.

Demographic variables include such measures as ethnicity, socioeconomic status, race, and age. Surveys often assess the preferences and attitudes of individuals, and many employ self-report scales to measure people's opinions and judgements about different items presented on a scale. Self-report scales are also used to examine the disparities among people on scale items.[4] These self-report scales, which are usually presented in questionnaire form, are one of the most used instruments in psychology, and thus it is important that the measures be constructed carefully, while also being reliable and valid.

Reliable measures of self-report are defined by their consistency. Thus, a reliable self-report measure produces consistent results every time it is executed. A test's reliability can be measured a few ways. First, one can calculate a test-retest reliability. A test-retest reliability entails conducting the same questionnaire to a large sample at two different times. For the questionnaire to be considered reliable, people in the sample do not have to score identically on each test, but rather their position in the score distribution should be similar for both the test and the retest. Self-report measures will generally be more reliable when they have many items measuring a construct. Furthermore, measurements will be more reliable when the factor being measured has greater variability among the individuals in the sample that are being tested. There will be greater reliability when instructions for the completion of the questionnaire are clear and when there are limited distractions in the testing environment. In contrast, a questionnaire is valid if what it measures is what it had originally planned to measure. Construct validity of a measure is the degree to which it measures the theoretical construct that it was originally supposed to measure.

It is important to note that there is evidence to suggest that self-report measures tend to be less accurate and reliable than alternative methods of assessing data (e.g. observational studies) Six steps can be employed to construct a questionnaire that will produce reliable and valid results:

1. one must decide what kind of information should be collected.
2. one must decide how to conduct the questionnaire
3. one must construct a first draft of the questionnaire.
4. the questionnaire should be revised

5. the questionnaire should be pretested
6. the questionnaire should be edited and the procedures for its use should be specified.

The way that a question is phrased can have a large impact on how a research participant answers the question. Thus, survey researchers must be conscious of their wording when writing survey questions. It is important for researchers to keep in mind that different individuals, cultures, and subcultures can interpret certain words and phrases differently from one another. There are two different types of questions that survey researchers use when writing a questionnaire:

- free response questions [questions are open-ended], and
- closed questions closed [usually multiple choice].

Free response questions are beneficial because they allow the responder greater flexibility, but they are also very difficult to record, tabulate and score, requiring extensive coding

Closed questions can be scored and coded more easily but they diminish expressivity and spontaneity of the responder. The vocabulary of the questions should be very simple and direct, and most should be less than twenty words. Each question should be edited for "readability" and should avoid 'leading' or 'loaded' questions.

Finally, if multiple items are being used to measure one construct, the wording of some of the items should be worded in the opposite direction to evade response bias [e.g. negatively posed questions included requiring a concerted effort to react to all questions in an even approach].

Surveyors must carefully construct the order of questions in a questionnaire. For questionnaires that are self-administered, the most interesting questions should be at the beginning of the questionnaire to catch the respondent's attention, while demographic questions should be near the end. However, if a survey is being administered over the telephone or in person, demographic questions should be administered at the beginning of the interview to boost the respondent's confidence. Question order may cause a survey response effect in which one question may affect how people respond to subsequent questions as a result of priming, and steps should be taken to inhibit this.

The following ways have been recommended for reducing nonresponse telephone and face-to-face surveys:

- Advance letter. A short letter is sent in advance to inform the sampled respondents about the upcoming survey. The style of the letter should be personalized but not overdone:

1. it announces that a phone call will be made, or an interviewer wants to make an appointment to do the survey face-to-face.
2. , the research topic will be described.
3. it allows both an expression of the surveyor's appreciation of cooperation and an opening to ask questions on the survey.

- Training. The interviewers are thoroughly trained in how to ask respondents questions, how to work with computers and making schedules for callbacks to respondents who were not reached.
- Short introduction. The interviewer should always start with a short introduction about him or herself. She/he should give her name, the institute she is working for, the length of the interview and goal of the interview. Also, it can be useful to make clear that you are not selling anything: this has been shown to lead to a slightly higher responding rate
- Respondent-friendly survey questionnaire. The questions asked must be clear, non-offensive and easy to respond to for the subjects under study.

Brevity is also often cited as increasing response rate. A 1996 literature review found mixed evidence to support this claim for both written and verbal surveys, concluding that other factors may often be more important.

Experimental studies

A common goal for a statistical research project is to investigate causality, and in particular to draw a conclusion on the effect of changes in the values of predictors or independent variables on dependent variables. There are two major types of causal statistical studies: experimental studies and observational studies. In both types of studies, the effect of differences of an independent variable (or variables) on the behaviour of the dependent variable are observed. The difference between the two types lies in how the study is actually conducted. Each can be very effective. An experimental study involves taking measurements of the system under study, manipulating the system, and then taking additional measurements using the same procedure to determine if the manipulation has modified the values of the measurements. In contrast, an observational study does not involve experimental manipulation. Instead, data are gathered and correlations between predictors and response are investigated. While the tools of data analysis work best on data from randomized studies, they are also applied to other kinds of data like natural experiments and observational studies for which a statistician would use a modified, more structured estimation method (e.g., Difference in differences estimation and instrumental variables, among many others) that produce consistent estimators.

The basic steps of a statistical experiment are:

1. Planning the research, including finding the number of replicates of the study, using the following information: preliminary estimates regarding the size of treatment effects, alternative hypotheses, and the estimated experimental variability. Consideration of the selection of experimental subjects and the ethics of research is necessary. Statisticians recommend that experiments compare (at least) one new treatment with a standard treatment or control, to allow an unbiased estimate of the difference in treatment effects.
2. Design of experiments, using blocking to reduce the influence of confounding variables, and randomized assignment of treatments to subjects to allow unbiased estimates of treatment effects and

experimental error. At this stage, the experimenters and statisticians write the *experimental protocol* that will guide the performance of the experiment and which specifies the *primary analysis* of the experimental data.
3. Performing the experiment following the experimental protocol and analyzing the data following the experimental protocol.
4. Further examining the data set in secondary analyses, to suggest new hypotheses for future study.
5. Documenting and presenting the results of the study.

The Hawthorne Effect

Experiments on [about] human behaviour have special concerns. The famous Hawthorne study examined changes to the working environment at the Hawthorne plant of the Western Electric Company. The researchers were interested in determining whether increased illumination would increase the productivity of the assembly line workers. The researchers first measured the productivity in the plant, then modified the illumination in an area of the plant and checked if the changes in illumination affected productivity. It turned out that productivity indeed improved (under the experimental conditions). However, the study is heavily criticized today for errors in experimental procedures, specifically for the lack of a control group and blindness.

The 'Hawthorne effect' refers to finding that an outcome (in this case, worker productivity) changed due to observation itself. Those in the Hawthorne study became more productive not because the lighting was changed but because they were being observed

Studies

An example of an observational study is one that explores the association between smoking and lung cancer. This type of study typically uses a survey to collect observations about the area of interest and then performs statistical analysis. In this case, the researchers would collect observations of both smokers and non-smokers, perhaps through a cohort study, and then look for the number of cases of lung cancer in each group. Case controlled is another type of observational study in which people with and without the outcome of interest (e.g. lung cancer) are invited to participate and their exposure histories are collected.

Various attempts have been made to produce a taxonomy of levels of measurement. The psychophysicist Stanley Smith Stevens defined four scales:

1. nominal,
2. ordinal,
3. interval, and
4. ratio

Other categorizations have been proposed. For example, Mosteller and Tukey defined:

1. grades,
2. ranks,
3. counted fractions,
4. counts,
5. amounts, and
6. balances.

Nelder (1990), Chrisman (1998), and van den Berg (1991) have also contributed to the categorisation schemes.

The issue of whether or not it is appropriate to apply different kinds of statistical methods to data obtained from different kinds of measurement procedures is complicated by issues concerning the transformation of variables and the precise interpretation of research questions. "The relationship between the data and what they describe merely reflects the fact that certain kinds of statistical statements may have truth values which are not invariant under some transformations. Whether or not a transformation is sensible to contemplate depends on the question one is trying to answer".

Most studies only sample part of a population, so results don't fully represent the whole population. Any estimates obtained from the sample only approximate the population value. Confidence intervals allow statisticians to express how closely the sample estimate matches the true value in the whole population. Often, they are expressed as 95% confidence intervals. Formally, a 95% confidence interval for a value is a range where, if the sampling and analysis were repeated under the same conditions (yielding a different dataset), the interval would include the true (population) value in 95% of all possible cases. This does *not* imply that the probability that the true value is in the confidence interval is 95%.

From the 'Frequentist' perspective, such a claim does not even make sense, as the true value is not a random variable. Either the true value is or is not within the given interval. However, it is true that, before any data are sampled and given a plan for how to construct the confidence interval, the probability is 95% that the yet-to-be-calculated interval will cover the true value: at this point, the limits of the interval are yet-to-be-observed random variables. One approach that does yield an interval that can be interpreted as having a given probability of containing the true value is to use a credible interval from Bayesian statistics: this approach depends on a different way of interpreting what is meant by "probability", that is, as a Bayesian probability.

Referring to statistical significance does not necessarily mean that the overall result is significant in real world terms. For example, in a large study of a drug it may be shown that the drug has a statistically significant but very small beneficial effect, such that the drug is unlikely to help the patient noticeably.

A p-value is the smallest significance level that allows the test to reject the null hypothesis. This is logically equivalent to saying that the p-value is the probability, assuming the null hypothesis is true, of observing a result at least as extreme as the test statistic. The smaller the p-value, the lower the probability of committing type I error.

Generationism

Harrisand other scholars identify the hypocrisy of older public opinion that's ready to disdain young people facing big social problems.' Many younger citizens are now bandying about [to each other] statistics to use in arguments with older people who are convinced they just aren't 'trying' hard enough to get jobs. Millennial backlash against the stereotyping of their so-called generation makes use of the same arguments against generational thinking that sociologists and historians have spent years developing. By drawing attention to the effects of the economic situation on their lives, pointing out that human experience isn't universal and predictable, and calling upon adults to abandon broad assessments in favour of specific understanding, they prove a point: that 'generational thinking confirms preconceived prejudices, but it's fatally flawed as a mode of understanding the world'.

Intergenerational Equity

The U.S. National Debt is often cited as an example of intergenerational inequity, as future generations will have the responsibility of paying it off. The U.S. National Debt has grown substantially over the past several decades. Relative to total GDP, the debt burden has worsened in the past several years. This is reflected in the graph format here:

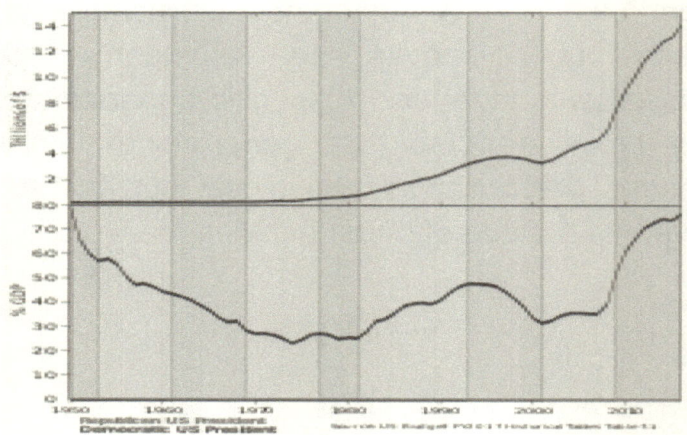

Intergenerational equity in economic, psychological, and sociological contexts, is the concept or idea of fairness or justice between generations. The concept can be applied to fairness in dynamics between children, youth, adults and seniors, in terms of treatment and interactions. It can also be applied to fairness between generations currently living and generations yet to be born, Study occurs across several fields. It is often discussed in public economics, especially with regard to:

- transition economics,
- social policy, and
- government budget-making.

Many cite the growing U.S. national debt as an example of intergenerational inequity, as future generations will shoulder 'consequences'. Intergenerational equity is also explored in environmental concerns [including:

- sustainable development,
- global warming and
- climate change.

The continued depletion of natural resources that has occurred in the past century will likely be a significant burden for future generations. Intergenerational equity is also discussed with regard to standards of living, with people of different ages and generations. Intergenerational equity issues also arise in the areas of elderly care and social justice.

Generationism

Intergenerational Standards of Living

Intergenerational equity in standards of living have shown differences between people of different ages as well as differences between people of different generations. Two perspectives on intergenerational equity in living standards have been distinguished by Rice, Temple and McDonald:

- a "cross-sectional" perspective – focuses on living standards at a particular point in time and how these living standards vary between people of different ages. The relevant issue is the degree to which, at a particular point in time, people of different ages enjoy equal living standards.
- a "cohort" perspective – focuses on living standards over a lifetime and how these living standards vary between people of different generations. For intergenerational equity, the relevant issue becomes the degree to which people of different generations enjoy equal living standards over their lifetimes.

Three indicators of intergenerational equity in economic flows, such as income, have been proposed by D'Albis, Badji, El Mekkaoui and Navaux. Their first indicator originates from a cross-sectional perspective and describes the relative situation of an age group (retirees) with respect to the situation of another age group (younger people). Their second indicator originates from a cohort perspective and compares the standards of living of successive generations at the same age. D'Albis, Badji, El Mekkaoui and Navaux's third indicator is a combination of the two previous criteria and is both an inter-age indicator and an intergenerational indicator.

In Australia, notable equality has been achieved in living standards, as measured by consumption, among people between the ages of 20 and 75 years. Substantial inequalities exist, however, between different generations, with older generations experiencing lower living standards in real terms at particular ages than younger generations. One way to illustrate these inequalities is to look at how long different generations took to achieve a level of consumption of $30,000 per year [In $ for what benchmark year?]. At one extreme, people born in 1935 achieved this level of consumption when they were roughly 50 years of age, on average. At the other extreme, Millennials born in 1995 had achieved this level of consumption by the time they were around 10 years of age. This has led some scholars to argue that standards of living have tended to increase generation over generation in most countries, as development and technology have progressed. It is also true that the 'value in purchasing power' of $30,000 has successively year-on-year diminished as a correlated of inflationary pressures] When taking this into account, younger generations may have inherent privileges over older generations, which may offset the redistribution of wealth towards older generations.

Adultism

Narrowly defined 'adultism is *prejudice and accompanying systematic discrimination against young people"*. On a more philosophical basis, the term has also been defined as *"bias towards adults... and the social addiction to adults, including their ideas, activities, and attitudes"*. The word adultism was used by Patterson Du Bois in 1903 and appears in French psychology literature in 1929, describing the influence of adults over children. It was seen as a condition wherein a child possessed adult-like "physique and spirit".

This definition was superseded by a late 1970s journal article proposing that adultism is the abuse of the power that adults have over children. The author identified examples of adultism not only in parents but in teachers, psychotherapists, the clergy, police, judges, and juries. Social science literature has identified adultism as "within the context of the social inequality and the oppression of children, where children are denied human rights and are disproportionately victims of maltreatment and exploitation."

The United Nations defines *youth* as persons between the ages of 15 and 24 with all UN statistics based on this range, the UN states education as a source for these statistics. The UN also recognizes that this varies without prejudice to other age groups listed by member states such as 18–30. A useful distinction within the UN itself can be made between teenagers (i.e. those between the ages of 13 and 19) and young adults (those between the ages of 18 and 32). The UN also states they are aware that several definitions exist for youth within UN entities such as Youth Habitat range of 15–32 and African Youth Charter range of 15–35.While seeking to impose some uniformity on statistical approaches, the UN itself is aware of contradictions between approaches in its own statutes. Hence under the 15–24 definition (introduced in 1981) children are defined as those under the age of 14 while under the 1979 Convention on the Rights of the Child, those under the age of 18 are regarded as children.

In Vietnam, widespread notions of youth are sociopolitical constructions for both sexes between the ages of 15 and 35. In Brazil, the term *youth* refers to people of both sexes from 15 to 29 years old.

The intergovernmental organisation Organisation for Economic Co-operation and Development [OECD] defines *youth* as "those between 15 and 29 years of age"

Accordingly, there is a dramatic range in what passes for youth: from 14 years at a base to a high of 35 years. That is more than many generation stereotype bounds suggest [e.g. a generation of 15 years].

Although linked to biological processes of development and aging, *youth* is also defined as a social position that reflects the meanings different cultures and societies give to individuals between childhood and adulthood. The term in itself when referred to in a manner of social position, can be ambiguous when applied to someone of an older age with very low social position; potentially when still dependent on their guardians. Scholars argue that age-based definitions have not been consistent across cultures [or times], and that it is more accurate to focus on social processes or benchmarks in the transition to adult independence when defining youth [e.g. marriage, employment].

Youth is the stage of constructing the '''self-concept'. This, for youth, is influenced by variables such as peers, lifestyle, gender, and culture. It is a time of a person's life when their choices are most likely to affect their future.

Adultism may be defined as the "behaviours and attitudes based on the assumptions that adults are better than young people, and entitled to act upon young people without agreement". It is also seen as, "an addiction to the attitudes, ideas, beliefs, and actions of adults. Adultism is popularly used to describe any discrimination against young people and is distinguished from ageism, which is simply prejudice on the grounds of age; not specifically against youth. Adultism is ostensibly caused by fear of children and youth. It has been suggested that 'adultism, which is associated with a view of the self that trades on rejecting and excluding child-subjectivity, has always been present in 'Western culture''.

Generationism

Fletcher suggests that adultism has three main expressions throughout society:

- Attitudinal Adultism: Personal feelings, assumptions, and beliefs that form a person's attitudes about young people. This is also called *internalized adultism*.
- Cultural Adultism: The shared attitudes, including beliefs and customs, promoting the assumption that adults are superior to anyone who is not identified as an adult, simply because of their age. This is also called *social adultism*.
- Structural Adultism: The normalization and legitimization of historical, cultural, institutional and interpersonal dynamics that routinely advantage adults while producing cumulative and chronic adverse outcomes for young people. This is also referred to as *institutional adultism*.

A study by the Crisis Prevention Institute of the prevalence of adultism found an increasing number of local youth-serving organisations addressing the issue. For instance, a local programme in Oakland, California, describes the impact of adultism, which "hinders the development of youth, in particular, their self-esteem and self-worth, ability to form positive relationships with caring adults, or even see adults as allies", on their website.

Three studied outcomes of adultism are:

- youth alienation,
- youth culture [shared desires and resulting affinity to other youths] and
- rebellion against more aged [and generally more powerful] biased groups.

Adultism is often used to describe the oppression of children and young people by adults, which is seen as having the same power dimension in the lives of young people as racism and sexism. It is treated as a generalization of paternalism, allowing for the broad force of adulthood beyond males, and may be witnessed in the infantalisation of children and youth. Pedophobia (the fear of children) and ephebiphobia (the fear of youth) have been proposed as the antecedents to adultism. Tokophobia, [the fear of childbirth], may also be a precursor; gerontophobia, or its antonym, gerontocracy, may be extensions of adultism.

At least one prominent organisation describes discrimination against youth as ageism, which is any form of discrimination against anyone due to their age. The National Youth Rights Association argues that ageism is a more natural and understandable term than adultism and thus is more commonly used among the young people affected by this discrimination. Advocates of using 'ageism' also believe it makes common cause with older people fighting against their own form of age discrimination.

Underpinnings of Adultism

In his seminal 1978 article, Flasher argues that adultism is born of the belief that 'children' are inferior, professing that adultism can be manifested as excessive nurturing, possessiveness, or over-restrictiveness, all of which are consciously or unconsciously geared toward excessive control of a child. Recently, theologians Heather Eaton and Matthew Fox proposed, "Adultism derives from adults repressing the inner child".

A 2006/2007 survey conducted by the Children's Rights Alliance for England and the National Children's Bureau asked 4,060 children and young people whether they have ever been treated unfairly based on various criteria (race, age, sex, sexual orientation, etc.). A total of 43%

of British youth surveyed reported experiencing discrimination based on their age, substantially more than other categories of discrimination like sex (27%), race (11%), or sexual orientation (6%).

Several scholastic works have identified multiple forms of adultism, offering a typology that includes:

1. internalized adultism In a publication of the W. K. Kellogg Foundation, University of Michigan Barry Checkoway asserts that internalized adultism causes youth to "question their own legitimacy, doubt their ability to make a difference" and perpetuate a "culture of silence" among young people. "Adultism convinces us as children that children don't really count," reports an investigative study, and it "becomes extremely important to us [children] to have the approval of adults and be 'in good' with them, even if it means betraying our fellow children. This aspect of internalized adultism leads to such phenomena as tattling on our siblings or being the 'teacher's pet,' to name just two examples." Other examples of internalized adultism include many forms of violence imposed upon children and youth by adults who are

reliving the violence they faced as young people, such as corporal punishment, sexual abuse, verbal abuse, and community incidents that include store policies prohibiting youth from visiting shops without adults, and police, teachers, or parents chasing young people from areas without just cause.

2. institutionalized adultism - Institutional adultism may be apparent in any instance of systemic bias, where formalized limitations or demands are placed on people simply because of their young age. Policies, laws, rules, organisational structures, and systematic procedures each serve as mechanisms to leverage, perpetuate, and instill adultism throughout society. These limitations are often reinforced through physical force, coercion or police actions and are often seen as double-standards. This treatment is increasingly seen as a form of gerontocracy. Institutions perpetuating adultism may include the fiduciary, legal, educational, communal, religious, and governmental sectors of a community. Institutional adultism may be apparent in any instance of systemic bias, where formalized limitations or demands are

placed on people simply because of their young age. Policies, laws, rules, organisational structures, and systematic procedures each serve as mechanisms to leverage, perpetuate, and instill adultism throughout society. These limitations are often reinforced through physical force, coercion or police actions and are often seen as double-standards. This treatment is increasingly seen as a form of gerontocracy. Institutions perpetuating adultism may include the fiduciary, legal, educational, communal, religious, and governmental sectors of a community. Institutions perpetuating adultism may include the fiduciary, legal, educational, communal, religious, and governmental sectors of a community.

3. Cultural adultism - Cultural adultism is a much more ambiguous, yet much more prevalent, form of discrimination or intolerance towards youth. Any restriction or exploitation of people because of their young age, as opposed to their ability, comprehension, or capacity, may be said to be adultist. These restrictions are often attributed to euphemisms afforded to adults on the basis of age alone, such as "better judgment" or "the wisdom of age".

Generationism

Stratification

Discrimination against age is increasingly recognized as a form of bigotry in social and cultural settings around the world. An increasing number of social institutions are acknowledging the positions of children and teenagers as an oppressed minority group. Many youths are rallying against the adultist myths spread through mass media from the 1970s through the 1990s.

Research compiled from two sources (a Cornell University nationwide study, and a Harvard University study on youth) has shown that social stratification between age groups causes stereotyping and generalization; for instance, the media-perpetuated myth that all adolescents are immature, violent and rebellious. Opponents of adultism contend that this has led to growing number of youths, academics, researchers, and other adults rallying against adultism and ageism, such as organizing education programmes, protesting statements, and creating organisations devoted to publicizing the concept and addressing it.

Generationism

Simultaneously, research shows that young people who struggle against adultism within community organisations have a high rate of impact upon said agencies, as well as their peers, the adults who work with them, and the larger community to which the organisation belongs.

There may be many negative effects of adultism, including ephebiphobia and a growing generation gap. A reactive social response to adultism takes the form of the children's rights movement, led by young people who strike against being exploited for their labour. Numerous popular outlets are employed to strike out against adultism, particularly music and movies. Additionally, many youth-led social change efforts have inherently responded to adultism, particularly those associated with youth activism and student activism, each of which in their own respects have struggled with the effects of institutionalized and cultural adultism. It is of consequence that 'most' youth have lost 'student status' long before the expiry of youth 'thresholds' of 24-35 years of age.

Educator John Holt proposed that teaching adults about adultism is a vital step to addressing the effects of adultism, and at least one organisation and one curriculum do that. Several educators have created curricula that seek to teach youth about adultism, as well. Currently, organisations responding to the negative effects of adultism include the United Nations, which has conducted a great deal of research in addition to recognizing the need to counter adultism through policy and programmes. The CRC has particular Articles (5 and 12) which are specifically committed to combatting adultism. The international organisation Human Rights Watch has done the same.

Common practice accepts the engagement of youth voice and the formation of youth-adult partnerships as essential steps to resisting adultism.

Some ways to challenge adultism also include youth-led programming and participating in youth-led organisations. These are both ways of children stepping up and taking action to call out the bias towards adults. Youth-led programming allows the voices of the youth to be heard and taken into consideration. Taking control of their autonomy can help children take control of their sexuality, as well. Moving away from an adultist framework leads to moving away from the idea that children aren't capable of handling information about sex and their own sexuality. Accepting that children are ready to learn about themselves will decrease the amount of misinformation spread to them by their peers and allow them to receive accurate information from individuals educated on the topic.

In its milder form, "adultism is about the misuse of power and does not refer to the normal responsibilities of adults in relation to children". Therefore, "addressing adultism is not about reversing the power structure ... [or] completely eradicating it": rather, "shedding adultism involves a negotiation of decisions".

Age of Majority

The age of majority is the threshold of adulthood as recognized or declared in law. It is the moment when minors cease to be considered such and assume legal control over their persons, actions, and decisions, thus terminating the control and legal responsibilities of their parents or guardian over them. Most countries set the age of majority at 18. The word *majority* here refers to having greater years and being of full age as opposed to the state of being a minor. The law in a given jurisdiction may not actually use the term "age of majority". The term typically refers to a collection of laws bestowing the status of adulthood. The age of majority does not necessarily correspond to the mental or physical maturity of an individual.

Age of majority should not be confused with the following nine constructs:

1. age of maturity,
2. age of sexual consent,
3. marriageable age,
4. school-leaving age,
5. drinking age,
6. driving age,
7. voting age,
8. smoking age,
9. gambling age,

Although a person may attain the age of majority in a particular jurisdiction, they may still be subject to age-based restrictions regarding matters such as the right to vote or stand for elective office, act as a judge, and many others. Age of majority can be confused with the similar concept of the age of license, which also pertains to the threshold of adulthood but in a much broader and more abstract way. As a legal term of art, "license" means "permission", and it can implicate a legally enforceable right or privilege. Thus, an age of license is an age at which one has legal permission from government to do something. The age of majority, on the other hand, is legal *recognition* that one has grown into an adult. Age of majority pertains solely to the acquisition of the legal control over one's person, decisions and actions, and the correlative termination of the legal authority of the parents (or guardian(s), in lieu of parent(s)) over the child's person and affairs generally.

Many ages of license are correlated to the age of majority, but they are nonetheless legally distinct concepts. One need not have attained the age of majority to have permission to exercise certain rights and responsibilities. Some ages of license are actually higher than the age of majority. For example, to purchase alcoholic beverages, the age of license is 21 in all U.S. states. Voting ages throughout the world have seen recent rolling back to the enfranchisement of younger potential voters.

In almost all places, minors who are married are automatically 'emancipated'. Some places also do the same for minors who are in the armed forces or who have a certain degree or diploma. In many countries minors can be emancipated: depending on jurisdiction, this may happen through acts such as marriage, attaining economic self-sufficiency [e.g. paying taxes], obtaining an educational degree or diploma, or participating in a form of military service. In the United States, all states have some form of emancipation of minors.

Youth Rebellion

We begin by pointing out that there are [in any labelled generation] youth and adults. There is no fixed age for the initiation of rebellion, but it is frequently pegged as commencing with puberty and ending with adulthood. This latter term 'occurs' when the social and legal impediments to independent action have all been met. Physical separation from parents often introduces the youth to adult permission and obligation [e.g. going away to College, moving away from home etc.] There are two common types of rebellion:

- against socially fitting (rebellion of non-conformity)

- against adult authority (rebellion of non-compliance.)

In both types, rebellion attracts adult attention by offending it. The younger person in each sub-set emphatically asserts 'individuality from what parents like or independence of what parents want, and in each case succeeds in provoking their disapproval. This is why rebellion, which is simply behaviour that deliberately opposes the ruling norms or powers that be, has been given a good name by the younger person and a bad one by adults. The reason why parents usually dislike the perceived rebellion is not only that it creates more resistance to their job of providing structure, guidance, and supervision, but because rebellion can lead to serious kinds of harm.

Rebellion can cause young people to rebel against their own self-interests -- rejecting childhood interests, activities, and relationships that often supported their self-esteem. Some salient effects on social interfaces include:

- It can cause them to engage in self-defeating and self-destructive behaviour - refusing to do school work or even physically hurting themselves.

- It can cause them to experiment with high-risk excitement - accepting dares that as a child they would have refused.

- It can cause them to reject safe rules and restraints - letting impulse overrule judgment to dangerous effect.

- And it can cause them to injure valued relationships - pushing against those they care about and pushing them away.

Accordingly, youth rebellion is not simply a matter of parental aggravation; it is also a matter of concern. Although the young person thinks rebellion is an act of independence, it actually never is. It is an act of dependency. Rebellion causes the young person to depend self-definition and personal conduct on doing the opposite of what others desire. Redefinition of autonomy is key. Partnership with adults on projects select by youth improves autonomy and places the adult in a position of acquiescence. A principal antidote for rebellion is the true independence offered by creating and accepting a challenge - the young person deciding to do something hard with themselves for themselves in order to grow themselves. The teenager who finds a lot of challenges to engage with, and who has parents who support those challenges, doesn't need a lot of rebellion to transform or redefine him or herself. To what degree a young person needs to rebel varies widely across a 'spectrum of causes'., In "*Born to Rebel*" (1997), Frank Sulloway posits that later born children tend to rebel more than first born. Some of his reasoning is because they identify less with parents, do not want to be clones of the older child or children who went before, and give themselves more latitude to grow in non-traditional ways. So, parents may find later born children to be more rebellious.

Secondary Schooling

(Home-based)

During the late middle school and early high school years most rebellion is about creating needed differentiation to experiment with identity and needed opposition to gather power of self-determination. When parents feel hard-pressed by these acts of rebellion (breaking social rules, running with wilder friends, for example) they are best served by allowing natural consequences to occur and by repeatedly providing positive guidance. They do this by continually making statements about, and taking stands for, choices that support constructive growth. Each time they do so, they provide the young person a fresh choice point to cooperate with them. Particularly when rebellion pushes hardest, as it usually does in mid adolescence, it is the responsibility of parents to keep communicating a reference that will guide the young person down a constructive path of growing up. In the words of one veteran parent who had shepherded two adolescents through periods of high rebellion, "what it takes is the gentle pressure of positive direction relentlessly applied."

Many high school rebellions occur as a result of delayed adolescence, the young person dramatically rebelling at last to liberate himself or herself from childhood dependency on parental approval for always being the "good child, [note that only children are often slower to separate from parents because of strong attachment and protracted holding on by both sides]. In high school these young people, with graduation into more independence looming a year or two ahead, may need to initiate late stage rebellions to get the separation and differentiation and autonomy they need to undertake this next momentous step. At this older age, risk taking can be more dangerous, while parents miss the loss of closeness and compatibility with their son or daughter that they have enjoyed for so many years.

Rebellion starts in early adolescence with the young person resisting parental authority by saying: "You can't make me!" Rebellion ends in the last stage of adolescence, trial independence, with the young person resisting personal authority. Having unseated parental authority for leading her life and supplanted it with his/her personal authority, he/she finds himself/herself rebelling against it.

In this semi-independent state adults must let the consequences of the young person's resistant choices play out and not interfere. How to end this rebellion against self-interest and accept adult-peer leadership authority in life is the last challenge of youth. It must be met before their 'adulthood' can truly begin.

Rebellion occurs not just between cohorts defined as generation [e.g. baby-boomer 'adults' at odd with millennial 'youths'] but also within the cohort among those who have established independence which defines them as adults, and younger members of the cohort [generation] who cannot yet be defined as adult.

Voting

There is a significant difference in the expression of voting franchise among younger voters between the US (2016 Presidential Election) and the Canadian (42nd Parliamentary Election 2015).

Table 1

Generationism

US Voting 2016 by Age Cohort

Birth Years	% Voting
1988-1999	41%
1972-1987	57%
1953-1972	63%
1971 and earlier	78%

In Canada Voter Turnout is higher (less apathy) in all age categories than that experienced in the US. For the 2015 Parliamentary elections, the breakdown is shown in Table 2:

Table 2

Voter Turnout by Age Group 2015 General Elections	
Birth Cohort	2015 % Voting
1993-1999	57.1

Generationism

1983-1992	57.4
1973-1982	61.9
1963-1972	66.6
1953-1962	73.7
1943-1952	78.8
Pre 1943	67.4
All Canada	66.1

Generation-X

In one 2011, survey analyses from the Longitudinal Study of American Youth found Gen Xers to be "balanced, active and happy" in midlife (between Birth Years 1967 and 1987) and as achieving a work-life balance. The Longitudinal Study of Youth is an NIH-NIA funded study by the University of Michigan which has been studying Generation X since 1987. That study group reported that "mean level of happiness was 7.5 and the median (middle score) was 8. Only four percent of Generation X adults indicated a great deal of unhappiness (a score of three or lower). Twenty-nine percent of Generation X adults were very happy with a score of 9 or 10 on the scale of 0-10."

Gen X have more children per household than those 'generations immediately preceding them. For decades, the norm has been two children per family. Between 1995 and 2000, however, the rate of women having three or more children jumped 61% to 18.4%.

It has been surmised that happiness with balance may be defined as contentment which, of itself, is apathy towards re-direction in the social milieu. They become 'conservative' (small c). They are by in-large disinterested in re-writing the culture in which they thrive. They are preoccupied with working to support children, crippling debt; - and more often than not count their Baby Boom parents as dependents for whom they must also be financially responsible.

Major benchmark events occurring within this sectioned group included; AIDS, Fall of Berlin Wall, Cold War, Assassination of Robert Kennedy, Vietnam War, Pop Culture, Spiraling Divorce Rates, Invention of Personal Computer, Multiple Recessions, Homelessness, and the Introduction of the Internet.

In the United Kingdom, a 2016 study of over 2,500 office workers conducted by Workfront found that survey respondents of all ages selected those from Generation X as the hardest-working employees in today's workforce (chosen by 60% of those workers). The grouping was also ranked highest among fellow workers for having the strongest work ethic (chosen by 59.5%), being the most helpful (55.4%), the most skilled (54.5%), and the best troubleshooters/problem solvers (41.6%)

Studies done by the Pew Charitable Trusts, the American Enterprise Institute, the Brookings Institution, the Heritage Foundation and the Urban Institute challenged the notion that each generation would be better off than the one that preceded. A report titled *Economic Mobility: Is the American Dream Alive and Well?* Focused on the income of males born between April 1964 and March 1974. The 2007 study emphasized that this generation's men made less (by 12%) than their fathers had at that same age in 1974, thus reversing a historical trend. It concluded that per year increases in household income generated by fathers/sons have slowed (from an average of 0.9% to 0.3%), not even keeping pace with inflation Adding a second earner to the family, as with male income, spousal income trends are also downward.

Viacom International Media Networks and Viacom, based on over 12,000 respondents across 21 countries, reported:

- An unconventional approach to sex,
- Ties to friendship and family,
- Desire for flexibility and fulfillment at work and
- An absence of midlife crisis.

Pew Research, a nonpartisan American think tank, describes Generation X as "intermediary between Baby Boomers and Millennials. On multiple factors such as attitudes on political or social issues, educational attainment, and social media use they 'score high'.

Generation X, which built the bridge from analog to digital accounts for 55% of business startup initiators in the US. Exemplars such as Elon Musk (Tesla Motors, Space-X) are indicative of entrepreneurs from that generation. Consequently, they're knowledgeable in the tech sphere, much like Millennial "digital natives." They know how to use social media with ease [81% of them are on Facebook, Instagram, and Twitter], though they are perceived to use it for information acquisition and exchange rather than socialization.

Children previously left home permanently when school or job 'suggested' they establish themselves as independent householders, but now they're also coming back home in large numbers to Generation X parents. Adults in their 20s and 30s are imposing themselves on their parents at record or near-record levels. Instead of downsizing as they age, many parents from the Baby Boom cohort must *upsize* to house the burdensome return of their children in newly extended families.

48% of Generation X persons expect to provide primary care for their ageing parents.

Although they have purchasing power (slightly higher per capita income than their percent of population), they have <u>less wealth than their parents did</u> at their age This is in large part due to:

- lower wages and job transience
- <u>large college debts</u> and
- the cost of caring for their parents, grandparents (who are living longer due to medical advances) and children.

Birth Cohorts and US Voting Experience

The following table uses the year of birth on the x-axis and the party favoured on the y axis. It shows that at 2014 younger voters had not associated with the social idealism espoused by the Democratic Party and maintained similar affiliations as those of the immediately preceding Baby Boom Generation, even being more inclined to the Republican Party than were Baby Boomers. Millennials also favoured Republican Party candidates in a similar vein, though to a lesser extent.

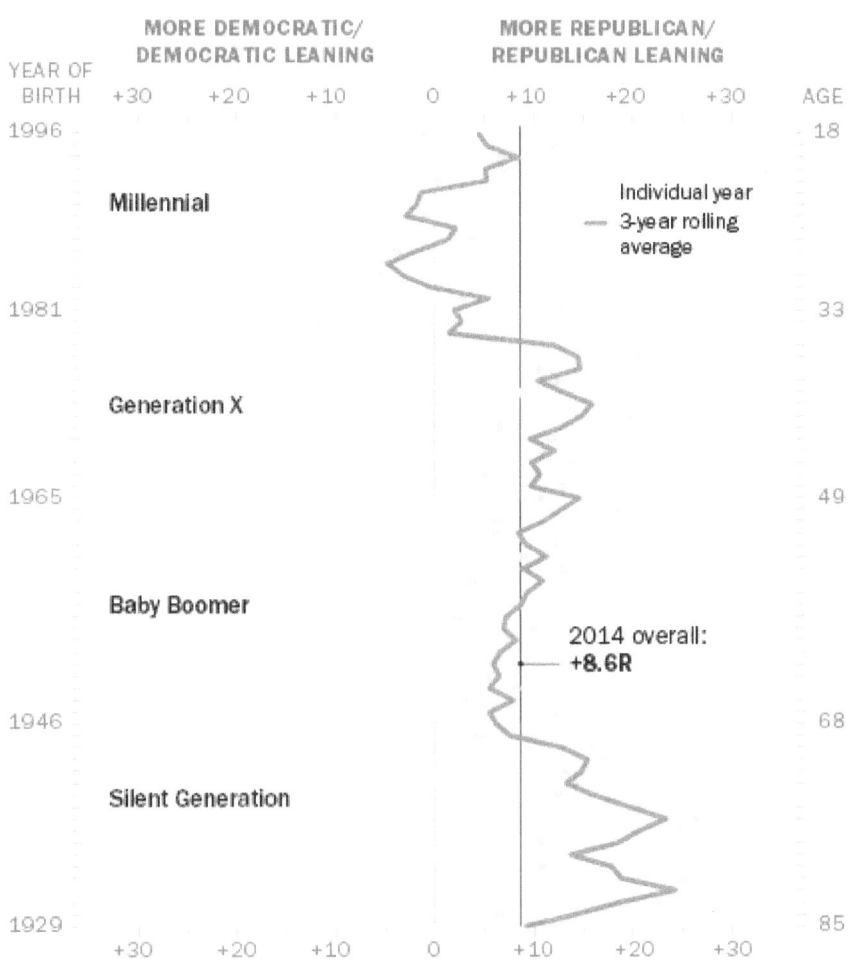

Generationism

Concluding Remarks

It has been demonstrated that there are significant It can be observed that Millennial and Generation X populations are less likely to vote than are Baby Boomer Generation populations, both in Canada and the US. Among Millennials there is a US rate of 41% voting while the equivalent Canadian rate is 57% - an increase in exercise of 39%. For Generation X the rates increase to 57% (US) and 67% (Canada). For Baby Boomers rates of voting are higher than either of the other later generations. In Australia 100% participation in voting is required of all generations of eligible voters (those male and female born before 1999). Failure to vote without legitimate excuse is a 'crime; in that country.

Generation X persons are self-identified as happy-contented with their lot, and accordingly do not especially engage themselves with social activism directed toward 'change' (as would be evidenced by exercise of voting franchise). This is suggestive of an apathy born of contentment with their lot, the status-quo and conservatism. It is demonstrated that this generation is hard-working to maintain their desired lifestyle, the rearing of children and support of parents. They change employers more frequently than prior generations to 'get ahead'. They also are more likely to act as entrepreneurs. Generation X workers are likely to be two partner working households as their individual incomes are considerably lower at the same ages than was the equivalent case for the preceding Baby Boomer Generation.

-30-

Selected Readings

- *Agresti, Alan; David B. Hichcock (2005). "Bayesian Inference for Categorical Data Analysis" (PDF). Statistical Methods & Applications.* **14** *(14): 298.* doi:*10.1007/s10260-005-0121-y*.
- Anderson, B.A.; Silver, B.D.; Abramson, P.R. (1988). "The effects of the race of the interviewer on race-related attitudes of black respondents in SRC/CPS national election studies". Public Opinion Quarterly. **52** (3): 1–28. doi:10.1086/269108.
- Anderson, D.R.; Sweeney, D.J.; Williams, T.A. (1994) *Introduction to Statistics: Concepts and Applications*, pp. 5–9. West Group. ISBN 978-0-314-03309-3
- *Drennan, Robert D. (2008). "Statistics in archaeology". In Pearsall, Deborah M. Encyclopedia of Archaeology. Elsevier Inc. pp. 2093–2100.* ISBN *978-0-12-373962-9*.
- Bogen, Karen (1996). "THE EFFECT OF QUESTIONNAIRE LENGTH ON RESPONSE RATES -- A REVIEW OF THE LITERATURE" (PDF). Proceedings of the Section on Survey Research Methods. American Statistical Association: 1020–1025. Retrieved 2013-03-19.

Generationism

- De Leeuw, E.D. (2001). "I am not selling anything: Experiments in telephone introductions". *Kwantitatieve Methoden*, 22, 41–48.
- Dillman, D.A. (1978) *Mail and telephone surveys: The total design method*. Wiley. ISBN 0-471-21555-4
- *Eisinga, R.; Te Grotenhuis, M.; Larsen, J.K.; Pelzer, B. (2011). "Interviewer BMI effects on under- and over-reporting of restrained eating. Evidence from a national Dutch face-to-face survey and a postal follow-up". International Journal of Public Health.* **57** *(3): 643–647. doi:10.1007/s00038-011-0323-z. PMC 3359459. PMID 22116390.*
- *Eisinga, R.; Te Grotenhuis, M.; Larsen, J.K.; Pelzer, B.; Van Strien, T. (2011). "BMI of interviewer effects". International Journal of Public Opinion Research.* **23** *(4): 530–543. doi:10.1093/ijpor/edr026.*
- *Everitt, Brian (1998). The Cambridge Dictionary of Statistics. Cambridge, UK New York: Cambridge University Press.* ISBN 0521593468.
- Fisher|1971|loc=Chapter II. The Principles of Experimentation, Illustrated by a Psycho-physical Experiment, Section 8. The Null Hypothesis

Generationism

- *Flores-Macias, F.; Lawson, C. (2008). "Effects of interviewer gender on survey responses: Findings from a household survey in Mexico". International Journal of Public Opinion Research.* **20** *(1): 100–110.* doi:*10.1093/ijpor/edn007*.
- Freedman, D.A. (2005) *Statistical Models: Theory and Practice*, Cambridge University Press. ISBN 978-0-521-67105-7
- *Freund, J. E. (1988). "Modern Elementary Statistics". Credo Reference.*
- *Galton, F (1877). "Typical laws of heredity". Nature.* **15** *(388): 492–553.* Bibcode:*1877Natur..15..492..* doi:*10.1038/015492a0*.
- George Beam. The Problem with Survey Research (2012) New Brunswick, NJ: Transaction, p. xv.
- *Groves, R.M.; Fowler, F. J.; Couper, M.P.; Lepkowski, J.M.; Singer, E.; Tourangeau, R. (2009). Survey Methodology. New Jersey: John Wiley & Sons.* ISBN 978-1-118-21134-2.
- Hand, D. J. (2004). *Measurement theory and practice: The world through quantification.* London: Arnold.
- Hays, William Lee, (1973) *Statistics for the Social Sciences*, Holt, Rinehart and Winston, p.xii, ISBN 978-0-03-077945-9

Generationism

- *Helen Mary Walker (1975). Studies in the history of statistical method. Arno Press.*
- *Hess, Gary. "American Tragedy; Kennedy, Johnson, and the Origins of the Vietnam War (review)". Johns Hopkins University. Project Muse. Retrieved 3 February 2017.*
- *Hill, M.E (2002). "Race of the interviewer and perception of skin color: Evidence from the multi-city study of urban inequality". American Sociological Review. **67** (1): 99–108. doi:10.2307/3088935. JSTOR 3088935.*
- Hoover, Eric "The Millennial Muddle". *The Chronicle of Higher Education*. October 11, 2009.
- *Howe, Neil (1992). Generations: The History of America's Future, 1584 to 2069. ISBN 978-0688119126.*
- *Howe, Neil (1992). Generations: The History of America's Future, 1584 to 2069. ISBN 978-0688119126.*
- Huff, Darrell (1954) *How to Lie with Statistics*, WW Norton & Company, Inc. New York, NY. ISBN 0-393-31072-8
- *Huff, Darrell; Irving Geis (1954). How to Lie with Statistics. New York: Norton. The dependability of a sample can be destroyed by [bias]... allow yourself some degree of skepticism.*

Generationism

- *Huff, Darrell; Irving Geis (1954). How to Lie with Statistics. New York: Norton.*
- *Ioannidis, J. P. A. (2005). "Why Most Published Research Findings Are False". PLoS Medicine. 2 (8): e124. doi:10.1371/journal.pmed.0020124. PMC 1182327. PMID 16060722.*
- *J Stigler, S. M. (1989). "Francis Galton's Account of the Invention of Correlation". Statistical Science. 4 (2): 73–79. doi:10.1214/ss/1177012580.*
- J. Franklin, The Science of Conjecture: Evidence and Probability before Pascal, Johns Hopkins Univ Pr 2002
- *Jones,, Gary L. (Fall 1992). "Strauss, William and Neil Howe 'Generations: The History of America's Future, 1584–2069' (Book Review)". Perspectives on Political Science. 21 (4): 218. ISSN 1045-7097. Retrieved 23 January 2012.*
- *Kaiser, David. "No End Save Victory: How FDR Led the Nation into War". Barnes & Noble. Retrieved 3 February 2017.*
- *Kane, E.W.; MacAulay, L.J. (1993). "Interviewer gender and gender attitudes". Public Opinion Quarterly. 57 (1): 1–28. doi:10.1086/269352.*
- *Lakshmikantham,, ed. by D. Kannan,... V. (2002). Handbook of stochastic analysis*

and applications. New York: M. Dekker. ISBN 0824706609.
- McCarney R, Warner J, Iliffe S, van Haselen R, Griffin M, Fisher P (2007). "The Hawthorne Effect: a randomised, controlled trial". *BMC Med Res Methodol.* **7** (1): 30. doi:10.1186/1471-2288-7-30. PMC 1936999. PMID 17608932.
- Mellenbergh, G.J. (2008). Chapter 9: Surveys. In H.J. Adèr & G.J. Mellenbergh (Eds.) (with contributions by D.J. Hand), Advising on Research Methods: A consultant's companion (pp. 183–209). Huizen, Johannes van Kessel Publishing.
- *Millennials: A profile of the Next Great Generation (DVD). WMFE & PBS.* ISBN 978-0-9712606-7-2.
- *Miller, Jon D.* "The Generation X Report: Active, Balanced, and Happy: These Young Americans are not bowling alone" *(PDF). University of Michigan, Longitudinal Study of American Youth, funded by the National Science Foundation.*
- *Moore, David (1992). "Teaching Statistics as a Respectable Subject". In F. Gordon and S. Gordon. Statistics for the Twenty-First Century. Washington, DC: The Mathematical Association of America. pp. 14–25.* ISBN 978-0-88385-078-7

- Moses, Lincoln E. (1986) *Think and Explain with Statistics*, Addison-Wesley, ISBN 978-0-201-15619-5. pp. 1–3
- Mosteller, F., & Tukey, J. W. (1977). *Data analysis and regression*. Boston: Addison-Wesley.
- Nelder, J. A. (1990). The knowledge needed to computerise the analysis and interpretation of statistical information. In *Expert systems and artificial intelligence: the need for information about data*. Library Association Report, London, March, 23–27.
- *Neyman, J (1934). "On the two different aspects of the representative method: The method of stratified sampling and the method of purposive selection". Journal of the Royal Statistical Society. **97** (4): 557–625. JSTOR 2342192*.
- Nikoletseas, M. M. (2014) "Statistics: Concepts and Examples." ISBN 978-1500815684
- *Pearson, K. (1900). "On the Criterion that a given System of Deviations from the Probable in the Case of a Correlated System of Variables is such that it can be reasonably supposed to have arisen from Random Sampling". Philosophical Magazine. Series 5. **50** (302): 157–175. doi:10.1080/14786440009463897*.

Generationism

- *Peterson, Peter G.; Neil Howe (1988). On Borrowed Time: How the Growth in Entitlement Spending Threatens America's Future. Google Books. ISBN 9781412829991. Retrieved 6 November 2012.*
- *Rothman, Kenneth J; Greenland, Sander; Lash, Timothy, eds. (2008). "7". Modern Epidemiology (3rd ed.). Lippincott Williams & Wilkins. p. 100.*
- *Rubin, Donald B.; Little, Roderick J. A., Statistical analysis with missing data, New York: Wiley 2002*
- *Schervish, Mark J. (1995). Theory of statistics (Corr. 2nd print. ed.). New York: Springer. ISBN 0387945466.*
- *Stanley, J. C. (1966). "The Influence of Fisher's "The Design of Experiments" on Educational Research Thirty Years Later". American Educational Research Journal. 3 (3): 223. doi:10.3102/00028312003003223.*
- *Stanley, Julian C. (1966). "The Influence of Fisher's "The Design of Experiments" on Educational Research Thirty Years Later". American Educational Research Journal. 3 (3): 223–229. doi:10.3102/00028312003003223. JSTOR 1161806.*

- *Strauss, William (1997). The Fourth Turning. Three Rivers Press.* ASIN B001RKFU4I.
- *Strauss, William (2009). The Fourth Turning. Three Rivers Press.* ASIN B001RKFU4I.
- van den Berg, G. (1991). *Choosing an analysis method.* Leiden: DSWO Press
- *Warne, R. Lazo; Ramos, T.; Ritter, N. (2012). "Statistical Methods Used in Gifted Education Journals, 2006–2010". Gifted Child Quarterly.* **56** *(3): 134–149.* doi:10.1177/0016986212444122.
- Willcox, Walter (1938) "The Founder of Statistics". *Review of the* International Statistical Institute 5(4): 321–328. JSTOR 1400906
- *Wolfram, Stephen (2002). A New Kind of Science. Wolfram Media, Inc. p. 1082.* ISBN 1-57955-008-8.
- *Zechmeister, E.; Jeanne, Z. (2011). Research methods in psychology (9th ed.). New York, NY:* McGraw Hill. *pp. 161–175.*
- *Giancola, Frank (2006). "The Generation Gap: More Myth than Reality". Human Resource Planning.* **29** *(4): 32–37.*
- *Brazhnikov, Pavel (2016).* "The theory of generations in the HR policy, the employers competition in the labor market" *(PDF). Trends and management.* **14** *(2): 194–201.*

Generationism

- Giancola 2006. "Some experts believe that the model is limited in its application to minorities and recent immigrants to North America (Robbins 2003); others have questioned its relevance to women (Quadagno, et al., 1993). "

- *Cohen, Jerome B. (December 1938). "Misuse of Statistics". Journal of the American Statistical Association. JSTOR. **33** (204): 657–674. doi:10.1080/01621459.1938.10502344.*
- *Yates, F (June 1964). "Sir Ronald Fisher and the Design of Experiments". Biometrics. **20** (2): 307–321. doi:10.2307/2528399. JSTOR 2528399.*
- *Box, JF (February 1980). "R. A. Fisher and the Design of Experiments, 1922–1926". The American Statistician. **34** (1): 1–7. doi:10.2307/2682986. JSTOR 2682986.*
- *Piccoli, Sean (3 April 1991). "13ers; The story of the new 'lost generation' (and America's hottest sound bite)". The Washington Times. p. E1.*
- *Bowman, James (April 5, 1991). "Another Grand Theory Comes of Age".*

- *Parshall, Gerald (8 April 1991). "History's Cycle Ride". U.S. News & World Report. Retrieved 21 October 2012.*
- *Alter, Jonathan (15 April 1991). "The Generation Game". Newsweek.*
- *Cormier, Jim (8 May 1993). "A young whine, with a sharp bite 13TH GEN: Abort, Retry, Ignore, Fail?". The Globe and Mail.*
- *Laurence, Charles (May 11, 1993). "The Bitter New Generation and Why They Are Criticizing Their Baby Boomer Parents". London Daily Telegraph.*
- *Leonard, Andrew (23 May 1993). "The Boomers' Babies". The New York Times. Retrieved 21 October 2012.*
- *Ferron, Alexander (July 1, 1993). "13th Generational Malaise". Eye Magazine.*
- *Kaiser, David (January 12, 1997). "Turning and turning in a widening gyre" (PDF). The Boston Sunday Globe. Retrieved 22 December 2012.*
- *Lind, Michael (January 26, 1997). "Generation Gaps". New York Times Review of Books. Retrieved 1 November 2010*
- *Brooks, David (5 November 2000). "What's the Matter With Kids Today? Not a Thing". The New York Times. Retrieved 30 January 2017.*

Generationism

- *Brooks, David (5 November 2000). "What's the Matter With Kids Today? Not a Thing". The New York Times. Retrieved 21 October 2012.*
- *Gomez, Dina (May 2001). "The next great generation" (PDF). NEA Today, V.19 No.4. Retrieved 22 December 2012.*
- *Gomez, Dina (May 2001). "The next great generation" (PDF). NEA Today, V.19 No.4. Retrieved 22 December 2012.*
- *Ringle, Ken (22 December 2007). "Bill Strauss: He Was the Life of the Parody". The Washington Post. Retrieved 24 May 2010.*
- *Hoover, Eric (11 October 2009). "The Millennial Muddle". The Chronicle of Higher Education. Retrieved 5 February 2017.*
- *Hoover, Eric (11 October 2009). "The Millennial Muddle: How stereotyping students became a thriving industry and a bundle of contradictions". The Chronicle of Higher Education. The Chronicle of Higher Education, Inc. Retrieved 11 January 2011.*
- *Horovitz, Bruce (May 4, 2012). "After Gen X, Millennials, what should next generation be?". USA Today.*
- *Howe, Neil (30 July 2014). "The G.I. Generation and the "Triumph of the Squares" (Part 2 of 7)". Forbes. Retrieved 10 February 2017.*

- *Howe, Neil (13 August 2014). "The Silent Generation, "The Lucky Few" (Part 3 of 7)". Forbes. Retrieved 10 February 2017.*
- *Howe, Neil (20 August 2014). "The Boom Generation, "What a Long Strange Trip" (Part 4 of 7)". Forbes.*
- *Howe, Neil (27 August 2014). "Generation X: Once Xtreme, Now Exhausted". Forbes*
- *Howe, Neil (4 September 2014). "The Millennial Generation, "Keep Calm and Carry On" (Part 6 of 7)". Forbes.*
- *Howe, Neil (October 27, 2014). "Introducing the Homeland Generation (Part 1 of 2)". Forbes. Retrieved 12 March 2015.*
- *Linette (2 February 2017). "Steve Bannon's obsession with a dark theory of history should be worrisome". Business Insider. Retrieved 3 February 2017.*
- Lynn, P. (2008) "The problem of non-response", chapter 3, 35-55, in *International Handbook of Survey Methodology* (ed.s Edith de Leeuw, Joop Hox & Don A. Dillman). Erlbaum. ISBN 0-8058-5753-2

About the Author

Mark Roberts-Seymour, B.A.Sc., P.Eng., CD, ACG, OFS is a Canadian Professional Forensic Engineer, a recognised stoneworks conservator, Chartered Demographer, non-fiction author, lay-brother, professional public speaker (Toastmaster Advanced Communicator Gold), technical paper referee, statistical assessor, and editor.

Over succeeding years his publications have included:

- *Old, Unemployed and Pissed: Late Career Canadians Coping with Long-term Unemployment*
- *Restructuring a Broken Canadian Economic-Democracy*
- *Three Proto-Christian Orthodoxies, The Gospel of Paul, Alexandrian Orthodoxy and Proto-Christian Gnosticism: A Comparison*

Generationism

- *Clinical Happiness: Measurement, Measures, Goals and Habit Conditioning*
- *Anglicanism: From Henry to Henrietta*
- *Cyborg:* Smartphone Reliance, AI and Transhumanism
- *An Afterlife:* Who Cares! Quotations on an After-life with Biographical notes
- *Conservation of Heritage Cemeteries*,
- *Green Revolutions – Will they be Enough*,
- *Suburban and Ex-Urban Self Sufficiency*,
- *Life Extension for Seniors Book 1: Expectations, Ageing Inhibition, Nutrition, Brain 'Wiring', Habits, Exercise and Ethics*
- *Old, Unemployed and Pissed: Late-Career Canadians Coping with Long-Term Unemployment*
- *The Working Poor: Who are Poor and What Can Be Done: A Central British Columbia Case Study*
- *Life Extension for Seniors Book 2: Methods, Research, Supplementation, Health and Life-Prolonging Strategies*
- *Abrahamic Gnosticism is not Scary*,
- *Proofs of God – Philosophy of Religion and Science Converge*,
- *Anglicanism – From Henry to Henrietta*,
- *Happiness: Biochemistry, Goals and Habituation*,
- *The Parable of the Prodigal Son - Death, Rebirth, Recognition and Reconciliation*,

Generationism

- *Nutrition for Older Workers*,
- *Nutrient Supplements for the Older Worker*,
- *Canadian Systems: Changing our Economic-Democracy*,
- *Gnosticism as Revelation: from St. Paul to C.G. Jung*,
- *Sabotage, Wealth and New Classes*,
- *Christian Metempsychosis: Elijah and John the Baptist*,
- *Sanctification – It's for Everyone!*,
- *The Gospel of Paul, Christian Gnosticism and Alexandrian Orthodoxy – A Comparison*,
- *An Afterlife, Who Cares! Quotations from 231 sources*
- *Renovating the Canadian Economic-Democratic System*,
- *To Coin a Phrase or Not to Coin a Phrase: Clichés, Metaphors and Euphemisms in Use*,
- *Canadian Systems: Changing our Economic-Democracy*,
- *It Can't Happen Here: Future Mechanisation, Despair and Suicide*

To access further information on these books [or for purchase], links to the Distributor Amazon.com are indicated in the list preceding.

Generationism

Mark acted for more thirty years as a Registered Professional Engineer (P.Eng., PE, ing.), technical author and editor for: The Government of BC (Lands Forests and Water Resources), BH Levelton and Associates, Warnock Hersey Professional Services, Heritage Technologies Press, RM Hardy and Associates, G.W. Spratt Limited and others. Mark also owned and managed a private Materials Engineering firm for an additional ten years (Roberts Seymour and Associates Limited).

He remains active in several service, political and social justice organisations as well as maintaining his professional standings. He is married, the father of four adult children and many grandchildren, and calls Vernon, British Columbia, Canada his home. Mark Roberts-Seymour can be reached directly by email at merscanada@gmail.com and by telephone at (250) 306-0550. For keynote speaking and seminar leadership contact (250) 721-5683.